Compositional Data Analysis in Practice

CHAPMAN & HALL/CRC
Interdisciplinary Statistics Series
Series editors: N. Keiding, B.J.T. Morgan, C.K. Wikle, P. van der Heijden

For more information about this series, please visit: https://www.crcpress.com/go/ids

Compositional Data Analysis in Practice

By
Michael Greenacre

CRC Press
Taylor & Francis Group
Boca Raton London New York

CRC Press is an imprint of the
Taylor & Francis Group, an **informa** business

A CHAPMAN & HALL BOOK

CRC Press
Taylor & Francis Group
6000 Broken Sound Parkway NW, Suite 300
Boca Raton, FL 33487-2742

© 2019 by Taylor & Francis Group, LLC
CRC Press is an imprint of Taylor & Francis Group, an Informa business

No claim to original U.S. Government works

Printed on acid-free paper
Version Date: 20180601

International Standard Book Number-13: 978-1-138-31661-4 (Hardback)
International Standard Book Number-13: 978-1-138-31643-0 (Paperback)

Visit the Taylor & Francis Web site at
http://www.taylorandfrancis.com

and the CRC Press Web site at
http://www.crcpress.com

In memoriam:
John Aitchison and Paul Lewi,
who both had a clear vision of
compositional data analysis.

Contents

Preface

Some background

I first became interested in compositional data analysis when I attended a talk by John Aitchison in Girona, Catalonia, in the year 2000. It was one of those life-changing moments that comes to one by sheer good luck. The first slide of John's talk was a blank triangle — I sensed immediately that this talk was going to be interesting!

The first time I learnt about triangular coordinates and the simplex geometry of compositional data was about two decades before meeting Aitchison, in the book *Geological Factor Analysis*, by Karl Jöreskog and co-authors Klovan and Reyment (1978). This book contained a figure showing how trivariate proportions that add up to 1 lie exactly in a planar triangle joining the unit points on three axes, [1 0 0], [0 1 0] and [0 0 1], in three-dimensional space, for example the RGB colour triangle which mixes the three colours red, green and blue in proportions to make any colour.

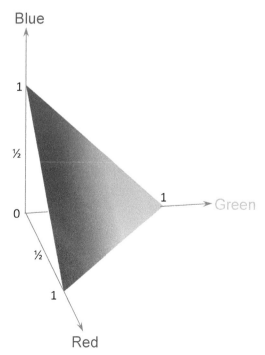

Since my doctoral studies in France in the 1970s, I had been involved with correspondence analysis, which analyses and visualizes sets of relative frequencies. So I had already realized the relevance of triangular (or ternary) coordinates for displaying trivariate proportions, and generalizations to a three-dimensional tetrahedron for quadrivariate proportions, and so on: in general, a hyper-simplex with n vertices in a space of dimensionality $n - 1$ for n-variate proportions.

I was working mainly with frequency data in the social and environmental sciences and had always considered frequencies relative to their total as the way to "relativize" the data — so I was always working in a simplex, specifically an irregular simplex structured by the chi-square distance. But John Aitchison's talk was about taking compositional data out of the simplex into unrestricted vector space. He did this by expressing data values as pairwise ratios, and then logarithmically transforming them, creating so-called "logratios". This was a revelation to me, and immediately after his talk, with great excitement, I expressed to John the potential of embedding his approach in a biplot of a compositional data matrix.

Of course, this was an obvious idea and not new — indeed, he told me that he had tried to publish an article 10 years earlier in the journal *Applied Statistics*, that it had been turned down and, out of irritation with the criticisms in the reports of the editor and referees, he had just dumped the whole thing. I was so interested that I asked him for that rejected article and the reports and then undertook a complete rewriting of the article. We eventually succeeded in getting it published by the same journal — see Aitchison and Greenacre (2002), listed in Appendix B. John's talk sparked my interest in this field, so closely allied to correspondence analysis. The publication of our joint paper completely cemented my interest and I have been actively working on compositional data analysis ever since.

Almost simultaneously with this immersion into compositional data analysis, I met Paul Lewi, the head of molecular biology at the company Janssen Pharmaceutica in Belgium. I had been aware of Lewi's work as early as the mid 1970s, having seen examples of what he called the *spectral map*, which looked just like a biplot. This was the visualization of a table of positive data that were first logarithmically transformed, then double-centred so that the row and column means were zero, and then this transformed table finally subjected to the usual dimension-reducing steps in forming a joint display of the rows and columns.

My first meeting with Paul, who came to a medical conference in Barcelona in 2002, rekindled my interest in his spectral map, and I soon realized that this method was almost identical to Aitchison and Greenacre's compositional biplot, with one crucial difference: the spectral map included weighting the rows and columns of the table proportional to the row and column marginal totals, just like in correspondence analysis. Paul and I subsequently published the connection between the spectral map, known since the 1970s but not well-known in the statistical literature, and the logratio biplot — see Greenacre and Lewi (2009) in Appendix B. I subsequently made another discovery, that the spectral map (called logratio analysis in this book) and correspondence analysis were not two completely unrelated methods, but actually part of the same family, linked by the Box-Cox power transformation that converges to the logarithmic function as the power parameter tends to zero.

As readers will learn in the present book, Paul's idea of weighting the rows and columns, both in computing averages as well as in the dimension reduction, is one of the most important aspects of compositional data analysis, a fact that Paul realized more than 40 years ago based on his extensive practical application of the spectral map to data on biological activity spectra. The default choice of weights, proportional to row and column margins, was actually inspired by Paul's knowledge of the work of Jean-Paul Benzécri, the originator of correspondence analysis. In 2005 I had the pleasure of visiting Paul and his colleagues at Janssen Pharmaceutica and participated in a Royal Statistical Society satellite meeting hosted by the company, and was amazed to see how the the spectral map had become one of the company's key analytical tools in drug development, leading to the discovery of drugs used in the treatment of many diseases such as polio and HIV/AIDS.

Aim and audience of this book

Having been immersed in compositional data analysis in recent years, and witnessing a tendency for the subject to drift into theoretical areas rather distant from the needs of practitioners, I decided to write this short book, with its primary aim to introduce an essentially practical approach to this widely applicable area of statistics. Because of this objective, the book is written for an audience of applied researchers in all the disciplines where compositional data are found: chemistry in general, mainly biochemistry and geochemistry, ecology, archaeology, political science, marketing, linguistics, sociology, psychology and the behavioural sciences, to name the principal areas of application. Statisticians will also find this an easy introduction to the subject, which echoes a talk given by John Aitchison at the Mathematical Geology conference in 1997, with the title: "The one-hour course in compositional data analysis, or compositional data analysis is simple". I do believe that compositional data analysis is simple, once one dominates the basic idea of a logratio, which is already a familiar statistical concept — just think of the logratio formed by the logit function, or logarithm of the odds, in logistic regression.

Structure of this book

The book follows a didactic format, originated in the second edition of my book *Correspondence Analysis in Practice* and continued into the third edition, which I have found to be excellent for teaching and learning: the ten chapters are exactly eight pages each, forming a "byte" of information that can be taught in 30-45 minutes, and able to be learnt in a concentrated hour of self-study.

Chapter 1 sets the scene with some simple examples, terminology and basic principles of compositional data analysis such as subcompositional coherence.

Chapter 2 explains the visualization of compositions and simplex geometry, as well as the concept of distance between compositions.

Chapter 3 deals with the fundamental transformation of compositional data using logratios. Several variants of the logratio transformation are defined and illustrated, including logratios involving amalgamations of compositional parts.

Chapter 4 is the most theoretically oriented chapter and can be skipped on a first reading. It deals with distributional properties of logratios, as well as the particular covariance structures engendered by the property that compositional values add up to a constant.

Chapter 5 shows various linear modelling exercises where compositional data are considered to be either explanatory variables or responses, always in the form of logratios. Total logratio variance is an important concept, since this is the total variance to be explained when compositional data comprise a set of responses. The multivariate technique of redundancy analysis is also described here.

Chapter 6 is concerned with reducing the dimensionality of compositional data, in the form of principal component biplots of logratios, specifically called logratio analysis. The weighting that was mentioned earlier is described here and weighted logratio analysis, which has many benefits, is compared to the unweighted form.

Chapter 7 shows how to cluster both the samples (rows) and parts (columns) of a compositional data set, generally using Ward clustering. Again, logratios of amalgamations are explored as a natural way of combining compositional parts.

Chapter 8 deals with one of the most problematic aspects of compositional data, the occurrence of zeros in the data, which prevent logratios from being computed. Several ways of handling this problem are discussed, for example substituting the zeros by justifiable small positive values, or using an alternative to logratio analysis such as correspondence analysis while checking on its lack of subcompositional coherence.

Chapter 9 is one of the most important chapters from a practical viewpoint, showing how to choose the most relevant logratios in a data set, depending on the objective of the study, or how to choose the most relevant subcompositions.

Finally, Chapter 10 gives a detailed case study of a data set in biology concerning the fatty acid compositions of several marine species, where the methods presented in this book are applied to compositional data with many samples and variables.

In addition, there are five appendices: Appendix A summarizes aspects of theory, Appendix B gives a commented bibliography, Appendix C shows computational examples using the R package easyCODA, Appendix D contains a glossary of terms, and finally I give some personal views in the form of an Epilogue.

Acknowledgments

This book is dedicated to the memories of John Aitchison and Paul Lewi, the two giants on whose shoulders we stand. I would like to thank Kimmo Vehkalahti who, during his sabbatical stay at the Universitat Pompeu Fabra in Barcelona, expertly proofread this manuscript. I am grateful to the specialists in fatty acid compositions, Martin Graeve, Stig Falk-Petersen, Janne Søreide, Anette Wold and Eva Leu, for their long collaboration, and to Professors Eric Grunsky, John Bacon-Shone, Trevor Hastie and Carles Cuadras, for their encouragement. Once again I thank Rob Calver and his team for their willingness and co-operation in publishing this book.

Michael Greenacre
Assos, July 2018

Chapter 1
What are compositional data, and why are they special?

Compositional data are a special case in the field of statistics. They are non-negative data with the distinctive property that their values sum to a constant, usually 1 or 100%. They are measures of parts (or components) of some total, where the total is usually of no interest. Their special character is due to the fact that their values depend on the particular choice of parts constituting the total amount. In this chapter compositional data are introduced and the use of ratios, preferably logarithmically transformed, is justified.

1.1 Examples of compositional data

Suppose you note down how much time you spend on a particular day sleeping, eating, travelling, working, relaxing, and so on, until you have accounted for the total of 24 hours — you will now have one set of compositional data. You can repeat this over a period of days, each day obtaining a different set of values quantifying your activities, until you have a table, or matrix, of these sets of observations, typically saved in a spreadsheet. The rows of this table will be the days, and the columns the different activities. The important aspect of these data is that each row adds up to 24 hours: this property of a constant sum for each set of values makes the data *compositional*. Why don't you try this out over a few weeks? If you do, you will have your own table of compositional data that you can analyse using the methods of this book, and so understand better your daily pattern of behaviour.

Now suppose you are a geologist involved in a prospection and studying the chemical properties of rocks. To do so, you need to analyse a sample of rock at each of several locations. The size of the rock depends on the instrumentation you use to analyse it, but the rock's size is irrelevant to your study — what you need to know is the *composition* of the rock sample, that is what proportion of the sample consists of elements such as silicon, sodium, potassium, calcium, magnesium, etc. These data are again compositional and are common in this field of geochemistry.

Marine biochemists also collect and analyse compositional data in the form of fatty acid proportions in the tissues of various marine organisms, from tiny plankton to huge whales. Fatty acids are passed up the marine food chain, and are also resynthesized along the way. Their relative amounts (i.e. compositions) in different species of fish, for example, as well as the compositional contents of their stomachs establish predator-prey relationships in the world's oceans and seas.

In economics, compositional data can be found in the form of household budgets, that is the proportions spent on different foodstuffs, education, travelling, entertainment, etc. Government budgets are also compositional in form, quantifying the proportions allocated to healthcare, education, defence, culture, and so on.

What about the food you eat, and its composition of protein, carbohydrates, fat and salt? On most bottled waters there are data on the various dissolved salts — in this case there might be some interest in the total amount of salts in the water (e.g. in parts per million) as well as their breakdown into components percentagewise.

This brings us to some areas of research where indeed it might well be the composition as well as the grand total of each sample that are of joint interest. Geographers studying land use might find that there is proportionally more agricultural land devoted to a specific crop where there is more land available. And political scientists studying the percentages of votes cast for political parties are also interested in how many votes were cast in total. Similarly, in community ecology as well as in botany, samples are taken of equal physical amounts (e.g. equal volumes in marine biology, or equal areas in botany, called "quadrats") and the species present in each sample are identified and counted. Ecologists are interested in both the composition of the samples, i.e. out of the total found in a sample, what proportion are of species A, B, C, etc., but at the same time whether the total found in the sample is related to the composition, for example that species A is proportionally higher in samples that are more abundant overall. Of additional interest in the same context is the diversity of each sample, e.g. how many different types of species are present, which also has a bearing on their compositional values.

In summary, compositional data are sets of values that add up to a fixed amount, usually 1 (proportions) or 100% (percentages). There are data that are compositional by their very nature, where the total size is completely irrelevant, as in the geochemical and biochemical examples. But there are other contexts where, in addition to the composition of the sample, it is also worth studying its "size", which is observed as part of the sampling process (e.g. the total amount of dissolved salts in water, how many fish are caught in a 15 minute trawl, how many people voted in an electoral district, how much land in total is allocated to agriculture).

1.2 Why are compositional data different from other types of data?

The fact that compositional data sum to a constant makes them special. In the more common situation when data are collected on several variables, for example in meteorology where temperature, pressure, humidity, etc., are measured, there is no constraint on such data to have any exact relationship, and each measurement is free to take a value on its particular measurement scale. Since percentages add up to 100%, however, their values are determined by the particular mix of components studied. For example, a nutritionist quantifying the compositions of various vegetables, might have the weights of protein, carbohydrate, fat and various other components, as percentages of the total combined weight of these nutritional components.

This gives one set of percentages, adding up to 100%, for each of the vegetables studied. Another nutritionist with a different angle on diet might include in the list the roughage contained in a vegetable, the non-nutritional part that is essential for a healthy diet. Now when this additional component is added, all the other percentages will be affected, decreased in this case. So a percentage value for protein, for example, will be determined by what other components are included in the set. This might seem an obvious point to make, and one might think that such percentages should always be expressed per fixed weight, say 100 g, of the vegetable, including its water and other components. But this has not deterred many researchers to report percentages and summary statistics such as means and standard deviations (std devns) when the percentages clearly depend on the choice of components included in the study.

A case in point is that of fatty acid analyses in marine biochemistry, mentioned in Sect. 1.1, where the set of fatty acids reported in one study is invariably different, but highly overlapping, with that of another study. This makes the reporting of summary statistics, for example averages, of each of the fatty acids rather meaningless, since each study is expressing the percentages with respect to totals that are not comparable. The example in Table 1.1(a) on the next page includes only four fatty acids, although there are usually from 25 to 40 fatty acids included in such a biochemical study, but it illustrates the problem and is based on real data. Table 1.1 shows, for a small sample of size 6, the compositional data for the four fatty acids[1], labelled *16:1(n-7)*, *20:5(n-3)*, *18:4(n-3)*, and *18:0* as a composition and then in Table 1.1(b) as a subcomposition, after dropping *18:0* and re-expressing the data as proportions. Clearly, the means and standard deviations change in the subcomposition: for example, in the full composition the mean of *16:1(n-7)* is 0.387 (std devn 0.233), while in the subcomposition it is 0.477 (std devn 0.228).

1.3 Basic terminology and notation in compositional data analysis

Every area of research has its own jargon and terminology, so here are some of the special terms that will be used throughout this book. At the same time, some mathematical notation will be introduced.

The components of a composition are called its *parts*. Parts will be indexed by the letter j and the number of parts, as well as the set of all parts, will be denoted by J. For example, suppose in the study of your own daily activities, the parts consist of sleeping, eating, travelling, working, relaxing and then all the other activities grouped together to compose a full day. This makes six parts, so $J = 6$, but also $J = \{$sleep, eat, travel, work, relax, other$\}$ — the alternative use of J will be clear from the context. This is an example of a six-part composition, in general a *J-part*

---

[1] Fatty acids, affectionately called the "Fats of Life", are abbreviated by their chemical structure. For example, *20:5(n-3)* has a chain of 20 carbon atoms with 5 double carbon bonds, and its last double bond is 3 carbons away from the methyl, or omega, end (i.e. it is an omega-3 fatty acid). See footnote on page 74 for more details.

Table 1.1 (a) Table of fatty acid compositions and (b) table of subcompositions after eliminating the last part. Means and standard deviations across the six samples are shown.

(a)

Samples	16:1(n-7)	20:5(n-3)	18:4(n-3)	18:0	Sum
B1	0.342	0.217	0.054	0.387	1
B2	0.240	0.196	0.050	0.515	1
D1	0.642	0.294	0.039	0.025	1
D2	0.713	0.228	0.020	0.040	1
H1	0.177	0.351	0.423	0.050	1
H2	0.209	0.221	0.511	0.059	1
Average	*0.387*	*0.251*	*0.183*	*0.179*	*1*
Std devn	*0.233*	*0.059*	*0.222*	*0.215*	

(b)

Samples	16:1(n-7)	20:5(n-3)	18:4(n-3)	Sum
B1	0.558	0.354	0.088	1
B2	0.494	0.403	0.103	1
D1	0.658	0.302	0.040	1
D2	0.742	0.237	0.021	1
H1	0.186	0.369	0.445	1
H2	0.222	0.235	0.543	1
Average	*0.477*	*0.317*	*0.207*	*1*
Std devn	*0.228*	*0.070*	*0.227*	

composition. The samples will be indexed by i and the total number of samples will be denoted by I, which can also denote the set of samples. For example, if you collect data for 30 days, during the month of April, 2018, then $I = 30$, but also I can stand for the set of days: $I = \{1/4/2018, 2/4/2018, \dots, 30/4/2018\}$.

What characterizes a set of compositional data is the constant sum constraint. Generally it is supposed that this constant sum is 1, even though the original data might be in hours, or counts, or parts per million, for example. The operation of dividing out a set of data by its total to obtain the compositional values, which are proportions adding up to 1, is called *closing* the data, or *closure*. Thus, having quantified the number of hours in the above six-part composition of daily activities, the data would be *closed* (i.e. divided by 24 in this case) to obtain values as proportions of the day. Some authors refer to *normalization* of the data as a synonym for closure, but this is preferably avoided since normalizing the data has another specific meaning in statistics.

If you were interested in activities during waking hours only, the hours of sleep would be dropped from the data set, new sample totals computed (i.e. total waking hours for each day), and the data *reclosed* to obtain new proportions of activities in the waking hours. Clearly, as already pointed out in Sect.1.2, the proportions will all change now that sleep has been excluded.

The five-part composition $J^* = \{$eat, travel, work, relax, other$\}$, excluding sleep, is called a *subcomposition* of the original six-part composition J defined above. The concept of a subcomposition is very important because in many cases the original data are essentially subcompositions. For example, in marine biochemistry it is impossible to identify every single fatty acid in a particular study, so the compositions studied are always subcompositions. It is the same situation in geochemistry, food research and many other contexts mentioned in Sect.1.1.

Mathematically, a compositional data set consists of a table of non-negative values with I rows (the samples, or other observational units) and J columns (the parts), denoted by a matrix $\mathbf{X}$ ($I \times J$), with general elements $x_{ij}, i = 1, \ldots, I, j = 1, \ldots, J$. The closure constraint is that the row sums are equal to 1: $\sum_j x_{ij} = 1$.

1.4 Basic principles of compositional data analysis

There are three very reasonable principles of compositional data analysis, which should be respected as faithfully as possible: scale invariance, subcompositional coherence and permutation invariance.

Scale invariance is a rather obvious property and is easily accepted. This principle states that compositional data carry only relative information, so that any change of the scale of the original data has no effect. So if the original data are multiplied by any scale factor C, for example a change of units, then the compositional data remain the same after closure.

Subcompositional coherence means that results obtained for a subset of parts of a composition, i.e. a subcomposition, should remain the same as in the composition. In Table 1.1 it was shown that computing the means and standard deviations of parts is not subcompositionally coherent. Computing correlations between parts has the same problem: for example, in Table 1.1(a) the correlation between *20:5(n-3)* and *18:4(n-3)* is 0.357, whereas in Table 1.1(b) it is −0.140 for the same pair of fatty acids.

Permutation invariance means that the results do not depend on the order that the parts appear in a composition. Of course, in a compositional data set the parts are all ordered in the same way for each sample, but the parts should be able to be re-ordered across the whole data set (i.e. columns of data set permuted) without affecting the results.

Both scale and permutation invariance appear as rather obvious and acceptable principles, whereas subcompositional coherence is the principle that will tend to dominate how to proceed when analysing compositional data.

An additional principle can be kept in mind, called *distributional equivalence*. This principle is one of the founding principles of correspondence analysis, which will be dealt with later (Chap. 8). Suppose that in your recorded data on daily activities, the time spent relaxing is always a fixed multiple of the time spent eating, say twice. These two parts are then considered to be distributionally equivalent and can be combined into one part, "relaxing and eating", without affecting the results of the analysis. Another situation might be in geochemistry when the percentage of an element A is always a fixed multiple of the percentage of another element B. Then A and B can be amalgamated, since they contain the same information, and the analysis of the data should not be affected at all.

1.5 Ratios and logratios

Subcompositional coherence dictates that results in a subcomposition must be the same as those when considering the larger composition. Referring back to Tables 1.1(a) and (b), we saw that the parts themselves are not subcompositionally coherent and give different results in compositions and subcompositions.

But now consider the ratio of fatty acids *16:1(n-7)/20:5(n-3)* in both tables, for the first sample: $0.342/0.217 = 1.576$ and $0.558/0.354 = 1.576$. These ratios are equal — in fact, all ratios are equal when comparing the two tables with respect to their common parts, and thus they are subcompositionally coherent.

The fundamental concept of compositional data analysis will thus be the ratio, which is invariant to the parts chosen. This means that two biochemists studying a different set of fatty acids can at least compare the ratios of two fatty acids that their studies have in common, and can summarize these ratios using regular statistical measures. To illustrate this equivalence, compare the three common columns in Tables 1.2(a) and (b) and check that they have the same values.

The ratios are all positive numbers and can have quite large and quite small values, as can be seen by inspecting Table 1.2. The standard deviations are usually

Table 1.2 Table of fatty acid ratios in (a) four-part compositions and (b) three-part subcompositions after eliminating the last part (see Table 1.1). The ratios common to both tables (first three columns) are identical. Means and standard deviations across the six samples are also shown.

(a)

Samples	16:1(n-7) / 20:5(n-3)	16:1(n-7) / 18:4(n-3)	20:5(n-3) / 18:4(n-3)	16:1(n-7) / 18:0	20:5(n-3) / 18:0	18:4(n-3) / 18:0
B6	1.576	6.333	4.019	0.884	0.561	0.140
B7	1.224	4.800	3.920	0.466	0.381	0.097
D4	2.184	16.462	7.538	25.680	11.760	1.560
D5	3.127	35.650	11.400	17.825	5.700	0.500
H5	0.504	0.418	0.830	3.540	7.020	8.460
H6	0.946	0.409	0.432	3.542	3.746	8.661
Average	*1.594*	*10.679*	*4.690*	*8.657*	*4.861*	*3.236*
Std devn	*0.943*	*13.573*	*4.176*	*10.523*	*4.307*	*4.158*

(b)

Samples	16:1(n-7) / 20:5(n-3)	16:1(n-7) / 18:4(n-3)	20:5(n-3) / 18:4(n-3)
B6	1.576	6.333	4.019
B7	1.224	4.800	3.920
D4	2.184	16.462	7.538
D5	3.127	35.650	11.400
H5	0.504	0.418	0.830
H6	0.946	0.409	0.432
Average	*1.594*	*10.679*	*4.690*
Std devn	*0.943*	*13.573*	*4.176*

as large or larger than the means, which indicates a high degree of skewness in their distributions. Remember that for the classic symmetric normal distribution 95% of the data are approximately contained in plus or minus two standard deviations around the mean. Hence, for the data in Table 1.2, two standard deviations below the mean go well below zero, which is impossible for positive numbers. For example, for the second ratio *16:1(n-7)/18:4(n-3)*, the mean plus or minus two standard deviations gives an interval $[-16.467, 37.825]$, which contains all six values of the ratio, but goes far into impossible negative values. The same problem is apparent in constructing confidence intervals for the mean, where these intervals often go into negative values as well. For this reason, and to make the data more symmetric, ratios are logarithmically transformed, converting strictly positive values into real numbers that can be negative or positive. But this is not the only justification for the log-transformation.

The logarithm is also used as a standard transformation of ratios of probabilities. For example, the *odds* of an event is defined as the probability p of the event happening divided by the complementary probability $1 - p$ of it not happening: i.e. odds $= p/(1 - p)$. For modelling purposes, the odds value is generally log-transformed to give the log-odds $= \log[p/(1 - p)]$ (this function, called the logit, will be discussed again in later chapters). The log-transformation is pervasive in statistics for converting *ratio-scale* data into *interval-scale* data. Ratios are generally compared multiplicatively, for example if the ratio $A/B = 1.2$, then it is considered twice the ratio $C/D = 0.6$, or ratio A/B is 100% more than ratio C/D. By log-transforming them, the ratios are linearized and the multiplicative difference becomes additive on an interval scale: $\log(A/B) = \log(2 \cdot C/D)$, i.e. $\log(A/B) = \log(2) + \log(C/D)$. If the ratios A/B and C/D are odds respectively, then $\log(2)$ is the logarithm of the ratio of the odds, or log-odds-ratio $\log[(A/B)/(C/D)] = \log[(AD)/(BC)]$.

Most statistical methods assume that data are on an interval scale, exemplified by the use of means (involving sums), variances (involving differences and sums), regression models (involving differences between observed and modelled values), etc. Hence, the log-transform is the key to convert ratios into the appropriate additive scale for statistical computations, as well as tending to symmetrize their distributions and reduce the effect of outliers.

Log-transformation of the ratios in Table 1.2(a) gives Table 1.3 (natural logarithms are always used as a standard). Fig. 1.1 plots the values of each ratio on the left and the log-transformed values on the right, showing how the distribution of each ratio is made more symmetric by the transformation. For example, the second ratio *16:1(n-7)/18:4(n-3)* has a clear outlier, which is diminished by the log-transformation.

Compositional data have been defined as being non-negative, not necessarily all positive. For ratios to be computed, it is clear that no data value can be zero. This raises a very tricky issue in compositional data analysis, which will be discussed in Chap. 8. Until then, only positive compositional data sets will be considered.

Table 1.3 Table of log-transformed ratios (i.e. logratios) of Table 1.2(a), as well as their means and standard deviations.

Samples	$\dfrac{16:1(n\text{-}7)}{20:5(n\text{-}3)}$	$\dfrac{16:1(n\text{-}7)}{18:4(n\text{-}3)}$	$\dfrac{20:5(n\text{-}3)}{18:4(n\text{-}3)}$	$\dfrac{16:1(n\text{-}7)}{18:0}$	$\dfrac{20:5(n\text{-}3)}{18:0}$	$\dfrac{18:4(n\text{-}3)}{18:0}$
B6	0.455	1.846	1.391	−0.124	−0.579	−1.969
B7	0.203	1.569	1.366	−0.764	−0.966	−2.332
D4	0.781	2.801	2.020	3.246	2.465	0.445
D5	1.140	3.574	2.434	2.881	1.740	−0.693
H5	−0.685	−0.871	−0.187	1.264	1.949	2.135
H6	−0.056	−0.894	−0.838	1.265	1.321	2.159
Mean	*0.306*	*1.337*	*1.031*	*1.295*	*0.988*	*−0.043*
Std devn	*0.643*	*1.861*	*1.278*	*1.586*	*1.418*	*1.960*

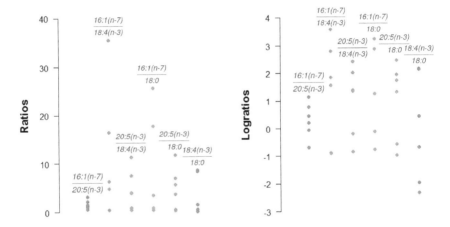

Fig. 1.1 *Dot plots of the ratios in Table 1.2(a) and logratios in Table 1.3.*

Summary: What are compositional data, and why are they special?

- Compositional data are nonnegative data on a set of *parts*, characterized by the sum constraint: the values of the parts from each observational unit add up to a fixed constant, usually 1 or 100%. The act of dividing a set of nonnegative data by its total to obtain a set of compositional data is called *closing*, or *closure*.
- Analysing the parts as regular statistical variables is an approach that is not *subcompositionally coherent* – that is, if a subcomposition (i.e. subset of parts) is analysed after closing the data again, results will be different compared to the original composition.
- Ratios of parts should rather be analysed, since they are subcompositionally coherent, remaining unchanged when analysed in subcompositions.
- Applying the logarithmic transformation converts ratios to an interval scale, reduces the effect of outliers and symmetrizes the distributions of the ratios. Hence, *logratios* will generally be used as the basic statistical variables for compositional data analysis.

Chapter 2
Geometry and visualization of compositional data

Knowing how to visualize compositional data is particularly helpful in understanding and interpreting their properties. The fixed sum constraint on compositional data leads to a special geometric representation of compositions in a space called a simplex. The simplest form of a simplex is a triangle, which contains three-part compositions. Four-part compositions are contained in a three-dimensional tetrahedron. Higher-dimensional simplexes, which cannot be visualized directly, contain compositions with more than four parts. In this chapter several ways of displaying compositional data are considered.

2.1 Simple graphics

Table 2.1 shows the data set `Vegetables` — these are compositions, as percentages of their respective totals, of protein, carbohydrate and fat in ten different vegetables. Confronted with a table of data such as this one, the casual data plotter might produce a figure such as Fig. 2.1. The vegetables are ordered exactly as in the original table, which was alphabetical, and the three percentages are plotted next to one another, with a legend at the bottom. There is nothing wrong with this plot, but there is not so much right about it as well — it does not take into account

Table 2.1 Data set `Vegetables`: Compositions, as percentages, of protein, carbohydrate and fat in 10 vegetables (Source: US Department of Agriculture, https://ndb.nal.usda.gov/ndb/nutrients/index).

Vegetables	*Protein*	*Carbohydrate*	*Fat*
Asparagus	35.66	61.07	3.27
Beans(soya)	42.05	35.88	22.07
Broccoli	48.69	43.78	7.53
Carrots	8.65	89.12	2.23
Corn	11.32	85.70	2.98
Mushrooms	16.78	77.25	5.97
Onions	10.44	88.61	0.95
Peas	26.74	71.29	1.97
Potatoes(boiled)	7.84	91.70	0.46
Spinach	41.57	52.76	5.67

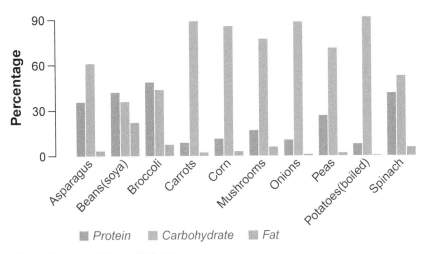

Fig. 2.1 Bar chart of data in Table 2.1.

that the data are compositional, and there is also no reason to maintain the original alphabetic ordering of the vegetables in the plot.

What is better is to plot the data as a compositional bar chart, as in Fig. 2.2, and to reorder the vegetables according to some feature of the data that assists the interpretation. In Fig. 2.2 the vegetables have been plotted as rows, so that the long labels can be positioned horizontally, and they have been ordered according to the carbohydrate percentages in the middle, in increasing order down the plot. Here it can be clearly seen that as carbohydrate percentages increase, the protein and fat percentages generally decrease accordingly, a consequence of the constraint that the percentages sum to 100%. Such a bar chart can be useful for compositions with

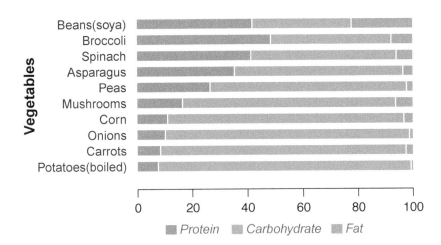

Fig. 2.2 Compositional bar chart of data in Table 2.1, plotted horizontally. The vegetables are ordered according to increasing percentages of carbohydrates.

Fig. 2.3 Scatterplot of two logratios, i.e. two ratios on logarithmic scales. The diagonal arrow, with slope −1 and increasing towards upper left, represents the logratio *Protein/Fat*. The projections of the vegetables perpendicularly onto this diagonal "axis" line them up in terms of this logratio, from lowest **Beans**) to highest (**Potatoes**), but with relatively low variance.

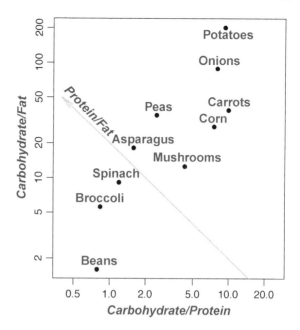

a modest number of parts, but the principle of subcompositional coherence should always be borne in mind – these percentages can change radically if additional parts are added. If data are expressed rather in parts per 100 grams of the vegetable, including the water and other components, a quite different pattern of percentages can emerge. For example, if water were included, the carbohydrate percentage for broccoli would be 2.88% compared to 11.24% for soya beans, due to the higher water content of broccoli, whereas in Table 2.1 and Fig. 2.2, broccoli is shown to have higher carbohydrate than soya beans, just the opposite!

For the three parts of Table 2.1 there are three possible ratios based on pairs of parts, giving logratios that can be plotted pairwise in scatterplots. Since any third logratio can be obtained linearly from the other two, there are only two "free" logratios, just as in the original data any third part can be obtained as 100 minus the sum of the other two parts. Fig. 2.3 shows the plot of the ratios *Carbohydrate/Protein* and *Carbohydrate/Fat*, using logarithmic scales on the two axes (i.e. logratios are plotted). In addition, an arrow with a slope −1 is shown which serves as an axis for the third logratio of Protein/Fat, because of the linear relationship:

$$\log(\textit{Protein/Fat}) = \log(\textit{Carbohydrate/Fat}) - \log(\textit{Carbohydrate/Protein}). \qquad (2.1)$$

If the vegetable points are projected perpendicularly onto this "axis", defined by the arrow, the values of $\log(\textit{Protein/Fat})$ can be recovered — all that is needed is to calibrate this diagonal axis on a logarithmic scale, like the two rectangular axes defining the scatterplot.

2.2 Geometry in a simplex

In Fig. 2.3, based on a three-part composition, it was necessary to choose two of the logratios to make the scatterplot using perpendicular axes, and the third logratio could be represented by a diagonal axis. In order to treat all parts equally, the original compositional values can be plotted in triangular coordinates, also known as ternary coordinates. A three-part composition can be represented exactly inside a triangle, thanks to the fixed sum constraint, and the triangle is the simplest form of what is called a *simplex*. Being familiar with the geometry of the simplex will also help to understand the geometry of correspondence analysis (Chap. 8).

A regular simplex, like the one shown in Fig. 2.4, has *vertices*, in this case three of them, at equal distances from one another. Each vertex represents a part of the composition. A composition is represented by a point inside the simplex. There are several ways to understand how a composition finds its position in the simplex, e.g. [0.42 0.36 0.22] for soya beans in Table 2.1 — notice that it will be preferable to think of compositions as values that sum to 1, and these proportional values have been rounded to two decimals for this explanation.

The first way is to think of a composition with respect to perpendicular axes, as shown in Fig. 2.4(a). All three-part compositions $[x \; y \; z]$ with the property that $x + y + z = 1$ will be contained in an equilateral triangle joining the unit points $[1 \; 0 \; 0]$, $[0 \; 1 \; 0]$ and $[0 \; 0 \; 1]$. So this triangle can effectively be "lifted out" of the three-dimensional space and laid flat to represent the compositions.

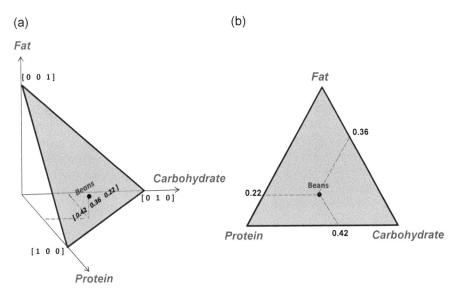

Fig. 2.4 (a) Three-part compositions of Table 2.1 plotted in three-dimensional Cartesian coordinates, with axis 1 = *Protein*, axis 2 = *Carbohydrate*, axis 3 = *Fat*. The vegetable **Beans** has coordinates [0.42 0.36 0.22] and lies inside the triangle linking the unit points on the axes. (b) The triangle can then be laid flat to show any three-part composition exactly.

The second way is to think of the sides of the triangle as axes representing the parts, calibrated in units of proportions: $0, 0.2, 0.4, \ldots, 1.0$ (see also Fig. 2.5 below). Then a composition can be situated in the triangle according to two out of its three values, e.g. 0.42 for *Protein* and 0.36 for *Carbohydrate*, and the third one (0.22) is redundant, as shown in Fig. 2.4(b). Notice that projections onto the side axes of the triangle are not done perpendicularly, as was done in Fig. 2.3, but parallel to a particular side of the triangle. These are called ternary axes, representing the three-part compositions in ternary coordinates.

Finally, a third way is to think of the vertices of the triangle as points on which weights are placed, equal to the values of the composition. The position of the composition is then at the weighted average. For example, the label Beans is positioned at the weighted average where a weight of 0.42 is placed on the vertex *Protein*, 0.36 on the vertex *Carbohydrate* and 0.22 on the vertex *Fat*. This is why its position is further away from the vertex *Fat* and slightly more towards the vertex *Protein* than to *Carbohydrate*. This way of thinking of the compositions as weighted averages, or *barycentres*, is perhaps the most intuitive — the points are also said to be plotted in *barycentric coordinates*, a term applicable to simplexes beyond two dimensions.

Whichever way you think about it, the positions of all ten vegetables in the simplex are shown in Fig. 2.5. Because most vegetables have high carbohydrate content in this three-part data set, their positions tend towards that vertex.

For four-part compositions and higher, additional dimensions are required to "see" the sampling units. The dimensionality of a composition is one less than its number of parts, so a four-part composition is three-dimensional and exists inside a simplex with four vertices — this is a tetrahedron and can still be visualized[1]. When

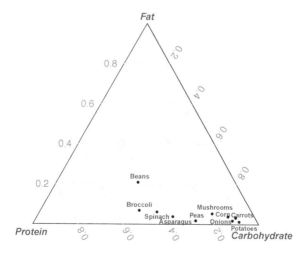

Fig. 2.5 Three-part compositions of Table 2.1 plotted in ternary coordinates inside a triangle, the simplest form of a simplex.

---

[1] A three-dimensional rotating view of a four-part composition is shown in the supplementary online material of this book — see Web Resources in Appendix B.

there are more than four parts, plots such as Figs 2.1 and 2.2 are still possible but additional technology will be required to visualize high-dimensional compositions (see Chap. 6).

2.3 Moving out of the simplex

Knowing the geometry of the simplex helps in understanding the original space of the compositions. The approach in compositional data analysis, however, is to rather move out of the simplex by converting the compositions into logratios and studying their structure. This was done in Fig. 2.3, for the three-part composition, but it was necessary to choose two of the three logratios as axes of the scatterplot.

If we use all three logratios as axes, knowing that one of them is exactly linearly related to the other two, the three logratios must themselves also be two-dimensional. Fig. 2.6 illustrates this for the vegetable data, where the three perpendicular axes are the logratios, the vegetable points are "suspended" in this space and lie exactly in the plane defined by Eqn (2.1). An additional property is that the plane goes through the origin (zero point) of the axes, because there is no constant value in the relationship of Eqn (2.1). This is the logratio space in which the compositional data will be visualized, rather than the simplex.

Because the compositional points lie exactly in a place, this plane can also be "laid flat" for interpreting the relative positions of the vegetables — see Fig. 2.7. This joint display of the samples (vegetables) and variables (logratios) is called a *biplot*. The way the logratio arrows are drawn is to make them emanate not from the original origins (zero points) of the three axes, but rather from the mean sample point, indicated by a small "+" at the centre of the vegetable points. This is equivalent to first centring the logratios with respect to their respective part means, which places the origin of the three-dimensional space at the means of all three logratios.

This is the simplest form of a biplot because it represents the data exactly. Later, in Chap. 6, biplots of higher-dimensional data will be treated, where it will be necessary to perform a dimension-reduction step in order to be able to view the higher-

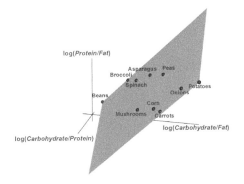

Fig. 2.6 Three-dimensional space of the three logratios. Since they are linearly dependent, defined by the relationship in Eqn (2.1), they lie on a two-dimensional plane, as shown. A three-dimensional rotation of this plot is given in the supplementary online material of this book.

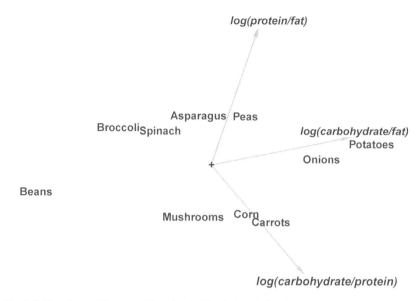

Fig. 2.7 The plane of the vegetable points in Fig. 2.6, with the three axes projected onto the plane to give a compositional biplot.

dimensional points approximately on a flat two-dimensional surface. The way to interpret the biplot of Fig. 2.7 is to project the vegetable points perpendicularly onto the three logratio axes — in this case their projected positions will be an exact display of their logratio values.

2.4 Distances between points in logratio space

Fig. 2.7 shows the positions of the 10 vegetables exactly in logratio space, but no scale is shown. What are the values of the distances between these points, for example between Asparagus and Beans, the first two rows of Table 2.1? This distance is the so-called *Euclidean distance* between the points using the three logratios as coordinates:

$$\sqrt{[\log \frac{61.07}{35.66} - \log \frac{35.88}{42.05}]^2 + [\log \frac{61.07}{3.27} - \log \frac{35.88}{22.07}]^2 + [\log \frac{35.66}{3.27} - \log \frac{42.05}{22.07}]^2}$$

$$\tag{2.2}$$

$$= 3.081$$

The more parts there are, the more this distance will increase in value. Hence, to keep the value "in check", as it were, it is convenient to divide the squared logratio differences by J^2, where J is the number of parts; in this example $J = 3$. Since this is a constant for all squared terms under the square root in (2.2), this is just $1/J = 1/3$ times the result above, namely $3.081/3 = 1.027$.

The general formula for this distance, called the *unweighted logratio distance*, between rows i and i' of $\mathbf{X}$, is thus:

$$d_{ii'} = \sqrt{\sum\sum_{j<j'} \frac{1}{J^2} \left[\log \frac{x_{ij}}{x_{ij'}} - \log \frac{x_{i'j}}{x_{i'j'}} \right]^2} = \frac{1}{J} \sqrt{\sum\sum_{j<j'} \left[\log \frac{x_{ij}\, x_{i'j'}}{x_{ij'}\, x_{i'j}} \right]^2} \quad (2.3)$$

where $\sum\sum_{j<j'}$ indicates the sum over all unique pairs (j, j') of columns (parts). The distances between all pairs of vegetables in Table 2.1 is given in Table 2.2. The largest distance is 1.966, between Potatoes and Beans, and the smallest distance 0.150, between Corn and Carrots. More general weighted distances, where different weights are assigned to the parts, will be considered later from Chap. 5 onwards.

Table 2.2 Unweighted logratio distances (2.3) between pairs of vegetables.

	Aspar	Beans	Brocc	Carro	Corn	Mushr	Onion	Peas	Potat
Beans (soya)	1.027								
Broccoli	0.477	0.589							
Carrots	0.735	1.373	1.050						
Corn	0.624	1.225	0.905	0.150					
Mushrooms	0.573	0.900	0.667	0.474	0.329				
Onions	0.756	1.658	1.203	0.452	0.530	0.826			
Peas	0.275	1.282	0.751	0.615	0.554	0.653	0.504		
Potatoes (boiled)	1.027	1.966	1.491	0.730	0.829	1.135	0.311	0.756	
Spinach	0.285	0.747	0.199	0.877	0.737	0.547	1.005	0.555	1.293

Summary: Geometry and visualization of compositional data

- Compositional data can be plotted in a compositional bar chart. If there are no prior orderings of the samples and the parts, these can be ordered in a way to facilitate the interpretation of the bar chart.
- Three-part compositions can be plotted in a triangle, which is the two-dimensional version of a simplex. The sides of the triangle can be considered as axes for the respective parts. Alternatively, the vertices of the triangle represent the parts and each composition is plotted at the weighted average of these corners, where the weights are the values of the parts.
- The logratio transformation takes the compositions out of the simplex into a Euclidean space. The dimensions of this space are the $\frac{1}{2}J(J-1)$ logratios of pairs of the J parts, but the samples lie in a subspace of dimensionality $J-1$, one less than the number of parts.
- Unweighted logratio distances between the samples can be computed between the samples, as Euclidean distances between their vectors of logratios.

Chapter 3
Logratio transformations

Logratios, i.e. log-transformed ratios, are the key to converting compositional data to scales that are additive and subcompositionally coherent. There are several variants of the simple logratio transformation that are worth considering for their practical and theoretical properties. In this chapter the additive logratio transformation is first considered, which gives a specific set of simple logratios. The set of centred logratios is mainly useful for facilitating several computations. The use of amalgamations in ratios is also considered, since amalgamations of parts are often used by practitioners of compositional data analysis. Finally, isometric logratios are treated, showing that they have interesting theoretical properties but are problematic in their substantive interpretation as statistical variables in practice.

3.1 Additive logratio transformations

The dimensionality of a J-part composition is equal to $J - 1$, one less than the number of parts. A set of independent logratios will be seen to completely describe and account for the variation in a compositional data matrix. By "independent" it is meant that any particular ratio in the set cannot be computed from the others: for example, the ratios A/C, A/B and B/C are not independent because $A/C = A/B \times B/C$. For logratios, this is the same as saying that the set of logratios should be linearly independent: for example, the logratios $\log(A/C)$, $\log(A/B)$ and $\log(B/C)$ are not independent because $\log(A/C) = \log(A/B) + \log(B/C)$.

The simplest logratio transformation, which provides a set of $J - 1$ independent logratios, is the *additive logratio transformation* (ALR), defined originally by John Aitchison. One part is chosen as the denominator of a set of of ratios defined by the other $J - 1$ parts as numerators. For example, denoting the parts by $X_1, X_2, \ldots, X_J$ and assuming the last part is chosen as the denominator, the ALRs are defined as follows:

$$\text{ALRs:} \quad \log\left(\frac{X_j}{X_J}\right) = \log(X_j) - \log(X_J), \quad j = 1, \ldots, J - 1 \tag{3.1}$$

The choice of the part for the denominator can be determined by the substantive nature of the data, or it can be decided according to a statistical criterion. For example, a nutritionist might use the carbohydrate proportion as the reference part for substantive reasons. In the absence of a natural reference part, the statistician could propose one that maximizes the concordance of the geometry of the ALRs with that of the complete set of pairwise logratios — this criterion, based on *Procrustes analysis*, will be considered in Chap. 8, Sect. 8.4. Whichever way the ALRs are defined, they are individually easy to understand and to interpret by the practitioner, and in this sense one of the most acceptable choices.

In matrix-vector notation, the full set of ALRs can be written as (cf. the formulation on the right hand side of (3.1)):

$$\text{ALRs:} \quad \begin{bmatrix} 1 & 0 & 0 & \cdots & 0 & -1 \\ 0 & 1 & 0 & \cdots & 0 & -1 \\ 0 & 0 & 1 & \cdots & 0 & -1 \\ \vdots & \vdots & \vdots & \ddots & \vdots & \vdots \\ 0 & 0 & 0 & \cdots & 1 & -1 \end{bmatrix} \begin{bmatrix} \log(X_1) \\ \log(X_2) \\ \log(X_3) \\ \vdots \\ \log(X_J) \end{bmatrix} \quad (3.2)$$

The coefficient matrix above has $J - 1$ rows and J columns.

3.2 Centred logratio transformations

A set of logratios that will be seen to have interesting theoretical properties and to be useful for purposes of computation is the set of *centred logratios* (CLRs). CLRs do not require choosing a reference part, but refer each of the J parts to the geometric mean of all the parts:

$$\text{CLRs:} \quad \log\left(\frac{X_j}{[\prod_j X_j]^{1/J}} \right) = \log(X_j) - \frac{1}{J} \sum_j \log(X_j), \quad j = 1, \ldots, J \quad (3.3)$$

i.e. CLRs are the log-transformed parts centred with respect to their mean across the parts. The fact that there are J CLRs automatically implies that they are linearly dependent: any one CLR can be computed from the $J - 1$ others, but any subset of $J - 1$ CLRs is linearly independent.

In matrix-vector notation the full set of CLRs can be written as follows:

$$\text{CLRs:} \quad \begin{bmatrix} 1 - \frac{1}{J} & -\frac{1}{J} & -\frac{1}{J} & \cdots & -\frac{1}{J} \\ -\frac{1}{J} & 1 - \frac{1}{J} & -\frac{1}{J} & \cdots & -\frac{1}{J} \\ \vdots & \vdots & \vdots & \ddots & \vdots \\ -\frac{1}{J} & -\frac{1}{J} & -\frac{1}{J} & \cdots & 1 - \frac{1}{J} \end{bmatrix} \begin{bmatrix} \log(X_1) \\ \log(X_2) \\ \vdots \\ \log(X_J) \end{bmatrix} \quad (3.4)$$

The coefficient matrix above has J rows and J columns. These are unweighted CLRs, where each part receives an equal weight $1/J$. Later, weighted CLRs will be preferred, where parts receive different weights (see Appendix A, (A.5), for the definition).

3.3 Logratios incorporating amalgamations

Compositional data often involve many parts and some parts form natural groups, based on substantive and/or statistical considerations. For example, a study of household food and drink purchases involves a multitude of items, many of which can be naturally amalgamated. Suppose alcohol purchases are recorded by types of alcohol: beer, wine, whiskey, and so on. If the simple ratios of these drinks are not of primary interest, these parts can be amalgamated into one part, called alcoholic beverages. Within this grouping, beer and wine could also be amalgamated for substantive reasons, being lower alcohol drinks, and ratioed relative to the amalgamation of spirits, and then finally the ratio of beer relative to wine might be a simple ratio of specific interest.

Another context is in the analysis of fatty acids in marine biochemistry, where amalgamations of polyunsaturated fatty acids (PUFA) and saturated fatty acids (SFA) are respectively formed, and the ratio PUFA/SFA computed for each individual studied. This ratio is also subcompositionally coherent in the sense that adding or removing any other fatty acid that is not part of these amalgamations leaves the ratio unaffected. But, of course, the adding or removing of any fatty acid within the PUFA or SFA groups will change the corresponding amalgamation and thus the ratio.

Comparing amalgamation ratios between studies thus needs to be conducted with care, and the compositions of the amalgamations should always be taken into account.

3.4 Isometric logratio transformations

Isometric logratios (ILRs) have been proposed as a theoretically more attractive way of combining subsets of parts, but as will be shown in several examples, they are complicated to interpret. They are thus not suitable as univariate statistics, and amalgamations might well be preferred. The definition of a single ILR is based on two subsets of parts, where a subset may contain just one part. Similar to a CLR the parts are combined by geometric means, and there is also a constant introduced into the definition, which further complicates its interpretation. Suppose J_1 denotes both the first subset of parts and the number of parts, and similarly J_2 for the second subset. Then the ILR defined on these two subsets is:

$$\text{ILR:} \quad \sqrt{\frac{J_1 J_2}{J_1 + J_2}} \log \frac{[\prod_{j \in J_1} X_j]^{1/J_1}}{[\prod_{j \in J_2} X_j]^{1/J_2}} = \sqrt{\frac{J_1 J_2}{J_1 + J_2}} \left(\frac{1}{J_1} \sum_{j \in J_1} \log(X_j) - \frac{1}{J_2} \sum_{j \in J_2} \log(X_j) \right)$$

(3.5)

The above definition is for a single ILR, and there have been several proposals to obtain a set of $J - 1$ ILRs to account for the full dimensionality of the compositional data set.

For example, one way of defining the subsets chosen for the numerator and denominator of each ILR is based on a hierarchical clustering of the parts. Omitting for the moment how this clustering is done (there are several possibilities, to be explained in Sect. 7.4), suppose that for a compositional data set on six daily activities (see Table 5.1), these activities cluster as in Fig. 3.1. There are five nodes in this dendrogram and each defines two subsets of activities and thus an ILR:

- {domestic} / {travel, free, work, sleep, meals}
- {travel, free} / {work, sleep, meals}
- {work} / {sleep, meals}
- {travel} / {free}
- {sleep} / {meals}

This set of ILRs is called a set of *balances*. For example, the balance for the first split above would be $\sqrt{5/6}$ times the logarithm of the ratio of time spent on domestic work divided by the geometric mean of the times for the other five activities.

Writing out the set of balances for the above set of splits on the six parts (X_1=domestic, X_2=travel, X_3=free, X_4=work, X_5=sleep, X_6=meals) gives the following matrix formulation (cf. version on right hand side of (3.5)):

$$
\text{ILRs:}
\begin{bmatrix}
\sqrt{\frac{5}{6}} & 0 & 0 & 0 & 0 \\
0 & \sqrt{\frac{6}{5}} & 0 & 0 & 0 \\
0 & 0 & \sqrt{\frac{2}{3}} & 0 & 0 \\
0 & 0 & 0 & \sqrt{\frac{1}{2}} & 0 \\
0 & 0 & 0 & 0 & \sqrt{\frac{1}{2}}
\end{bmatrix}
\begin{bmatrix}
1 & -\frac{1}{5} & -\frac{1}{5} & -\frac{1}{5} & -\frac{1}{5} & -\frac{1}{5} \\
0 & \frac{1}{2} & \frac{1}{2} & -\frac{1}{3} & -\frac{1}{3} & -\frac{1}{3} \\
0 & 0 & 0 & 1 & -\frac{1}{2} & -\frac{1}{2} \\
0 & 1 & -1 & 0 & 0 & 0 \\
0 & 0 & 0 & 0 & 1 & -1
\end{bmatrix}
\begin{bmatrix}
\log(X_1) \\
\log(X_2) \\
\log(X_3) \\
\log(X_4) \\
\log(X_5) \\
\log(X_6)
\end{bmatrix}
\quad (3.6)
$$

Notice that the rows of the coefficient matrix above have scalar products equal to zero, that is they are *orthogonal*. For example, the scalar product between the first two rows is $0 - \frac{1}{10} - \frac{1}{10} + \frac{1}{15} + \frac{1}{15} + \frac{1}{15} = 0$. The square root constant gives unit length (i.e. unit sum-of-squares) to each row. For example, for the first balance the sum-of-squares is $\frac{5}{6}(1 + 5 \times \frac{1}{25}) = \frac{5}{6} \times \frac{6}{5} = 1$. Hence, the coefficient matrix has *orthonormal* rows. This is an attractive theoretical property, and means that the ILRs form a set of *log-contrasts* with orthonormal coefficients What is gained theoretically, however, is lost in the substantive interpretation of the ILRs, as discussed in the next section.

Another type of ILR has been proposed, called *pivot logratios* (PLRs). These are a special case of balances and depend on the ordering of the parts. The numerator is

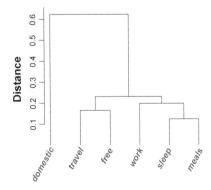

Fig. 3.1 Cluster dendrogram of time budgets. Activities: domestic work, travelling, free time, work, sleep and meals.

a single part and the denominator is the geometric mean of those parts "to the right" in the ordered list.

PLRs: $\sqrt{\dfrac{J-k}{J-k+1}} \log \dfrac{X_k}{[\prod_{j=k+1}^{J} X_j]^{1/(J-k)}}$

$$= \sqrt{\frac{J-k}{J-k+1}} \left(\log(X_k) - \frac{1}{J-k} \sum_{j=k+1}^{J} \log(X_j) \right), \quad k = 1, \ldots, J-1. \quad (3.7)$$

In matrix notation, the $J-1$ PLRs, which also form an orthonormal set, are:

PLRs:
$$\begin{bmatrix} \sqrt{\frac{J-1}{J}} & 0 & 0 & \cdots & 0 \\ 0 & \sqrt{\frac{J-2}{J-1}} & 0 & \cdots & 0 \\ 0 & 0 & \sqrt{\frac{J-3}{J-2}} & \cdots & 0 \\ \vdots & \vdots & \vdots & \ddots & \vdots \\ 0 & 0 & 0 & \cdots & \sqrt{\frac{1}{2}} \end{bmatrix} \begin{bmatrix} 1 & -\frac{1}{J-1} & -\frac{1}{J-1} & \cdots & -\frac{1}{J-1} & -\frac{1}{J-1} \\ 0 & 1 & -\frac{1}{J-2} & \cdots & -\frac{1}{J-2} & -\frac{1}{J-2} \\ 0 & 0 & 1 & \cdots & -\frac{1}{J-3} & -\frac{1}{J-3} \\ \vdots & \vdots & \vdots & \ddots & \vdots & \vdots \\ 0 & 0 & 0 & \cdots & 1 & -1 \end{bmatrix} \begin{bmatrix} \log(X_1) \\ \log(X_2) \\ \log(X_3) \\ \vdots \\ \log(X_J) \end{bmatrix}$$

(3.8)

There are very many ways of creating ILR balances or ordering the parts in the case of PLRs, and each will define a different set of variables. Like CLRs, ILRs can incorporate weights that are assigned to the parts — see Appendix A, (A.9).

3.5 Comparison of logratios in practice

Recall that a J-part composition has dimensionality one less, $J-1$. Furthermore, (unweighted) logratio distances between samples are computed by (2.3). For the three-part compositions of the vegetable data in Table 2.1, several options for selecting two variables in the form of ratios are considered in Fig. 3.2.

First, assuming that the researcher is interested primarily in the fat content, and so expresses the component *Fat* as a ratio of what is not fat (i.e. *Protein+Carbohydrate*) and then, within the amalgamation, the ratio of *Protein* to *Carbohydrate*. The two variables define the axes of the scatterplot in Fig. 3.2(a), with the amalgamation ratio along the horizontal axis showing the steadily increasing fat content. The second axis splits the vegetables into two sets rising in almost equal diagonal trends from lower left to upper right, with the upper set (from Potatoes to Broccoli) showing higher ratios of protein versus carbohydrate, for a given fat proportion. The interpoint distances in this plot are not exactly the logratio distances, but they are very close to them, as shown in the distance plot alongside. The Pearson correlation measuring how close the two sets of distances are to a straight line is equal to 0.9890.

Second, the same style of plot is shown in Fig. 3.2(b), but using the geometric mean, not the amalgamation, in the denominator of the first ratio, in other words the two ratios now form a set of balances. The configuration is hardly different from the previous one but now, according to theory, the distances are exactly the same as the logratio distances, apart from the scaling factor which is of no consequence. The correlation between the distances is equal to 1, as expected from theory. It is this

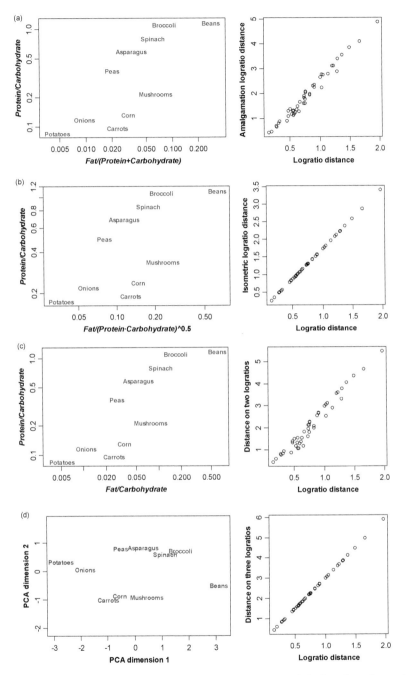

Fig. 3.2 Plots of vegetables according to their protein, fat and carbohydrate logratios, and concordance of the ensuing two-dimensional distances with the logratio distances: (a) Amalgamation logratios; (b) Isometric logratios; (c) Two additive logratios; (d) PCA of all three logratios. Notice the logarithmic scales in the left hand side plots of (a)–(c).

exact agreement that is claimed to be the benefit of using ILRs, but it can be seen that ratios using amalgamations or just simple logratios approximate the data structure very closely. However, it is the interpretation of ILRs as substantive variables, when they involve geometric means, that is problematic — see the next section.

Third, two simple logratios can be selected, *Fat* vs. *Carbohydrate* and *Protein* vs. *Carbohydrate* (i.e. the ALRs with respect to *Carbohydrate*). The plot in Fig. 3.2(c) is again similar to the others, with the two ascending subsets of points but slightly less separated. The correlation between the distances is 0.9814.

Finally, all three logratios are used, and the planar view of the vegetables is extracted, as was done in the biplot of Fig. 2.7 (here the ratio vectors have been omitted and the plot's horizontal axis has been reversed so that the plot agrees more with the configurations above). This configuration, obtained by principal component analysis, is identical to that of Fig. 3.2(b), apart from the rotation of the points, and the correlation is again equal to 1. Notice that the two subsets of points, lying in a diagonal orientation in Figs 3.2(a)–(c), are now neatly following the first principal axis, and their separation is defined by the second axis.

3.6 Practical interpretation of logratios

A simple ratio is clearly the easiest transformation to understand. The logarithm of this ratio does change the scale, but it is a standard transformation and easy to back-transform using the exponential function. Using amalgamations in the ratios is also easy for practitioners to digest, since a sum of parts can be considered like any other part, with an understandable labelling and interpretation.

However, ILRs are not easy to interpret, and even in this simple case of three parts, their substantive meanings are not clear. The same holds for CLRs, which are useful for computational purposes, as will be shown later, but not as variables for describing a composition. For example, would there be any worth in comparing the same ILR or the same CLR in two different studies?

In the two plots of Fig. 3.3, for a constant *Fat* value of 5% (0.05), the red dashed lines show the constant values of the ratio *Fat/(Protein+Carbohydrate)*, $0.05/0.95 = 0.053$ (analogous to the "odds" of *Fat*), where the right hand plot shows the corresponding compositions in the ternary diagram. In the left hand plot the black curve shows how the ratio in the ILR can vary for the same fixed value of 0.05 for *Fat*. Depending on the value of *Carbohydrate* (and thus of *Protein* by implication), the ILR can be seen to be as low as the log of $0.05/0.475 = 0.105$, when *Protein = Carbohydrate* = 0.475, or as high as the log of 0.50 for extreme values of *Protein* and *Carbohydrate*. The value of an ILR depends on the relative values of the parts in the geometric averages that form the ILR, and this makes it a very tricky statistic to interpret and to compare between studies.

In Fig. 3.3(b) the two curves trace out possible compositions corresponding to an ILR and a CLR, respectively, with a fixed value $\log(0.5)$. Clearly, there are very many different compositions that give the same value, both for ILRs and CLRs. For example, the same value of an ILR can correspond to a *Fat* percentage as high as 20% and as low as 5%.

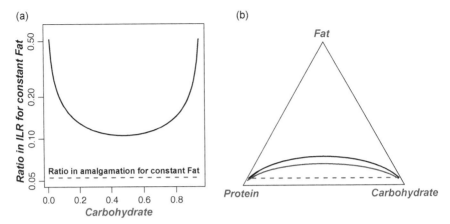

Fig. 3.3 (a) For a *Fat* value of 0.05 (5%), the amalgamation logratio of *Fat/(Carbohydrate + Protein)* is the logarithm of $0.05/0.95 = 1/19 = 0.053$, a constant shown by the red dashed line (note the logarithmic scale vertically). On the other hand the black curve shows the ratio in the ILR, $0.05/(\sqrt{Protein \times Carbohydrate})$ that varies depending on the product in the denominator, which changes for different values of *Carbohydrate*, and thus also of *Protein*. (b) In the ternary space of the compositions, the constant amalgamation ratio corresponds to a constant value of *Fat*, tracing out the red dashed line in the triangle, whereas the black curve shows points in the triangle corresponding to a fixed value of ILR, set at $\log(1/2)$, i.e. the ratio in the logarithm $= 0.5$. The blue curve traces out all CLRs that have constant values $\log(0.5)$.

Summary: Logratio transformations

- Logratios are the log-transforms of ratios of single parts or combinations of parts.
- Additive logratios (ALRs) express each of the $J - 1$ parts relative to the J-th one, where the part for the denominator can be any one of the parts.
- Centred logratios (CLRs) express each part with respect to the geometric mean of all the parts.
- Isometric logratios (ILRs) express the geometric means of two subsets of parts relative to each other. They also have a normalizing constant that makes the sum-of-squares of the coefficients of the part logarithms equal to 1.
- A special case of a set of ILRs, called balances, involves pairs of subsets of parts obtained from a recursive binary partitioning of the parts.
- Logratios can be formed from amalgamations of parts
- The logratios that have the most benefit in terms of practical interpretation are simple logratios of parts, including ALRs, and amalgamation logratios. Isometric and centred logratios have interesting theoretical properties when all pairwise ratios are of interest, but are difficult to interpret substantively.
- Centred logratios are very useful for computational purposes (to be demonstrated in subsequent chapters).

Chapter 4
Properties and distributions of logratios

One can say that logratio transformations bring compositional data from the "simplex world" into the "real world", that is into an unbounded space of real numbers where classical statistics can be applied. One consequence is that logratios might follow the normal distribution, the univariate normal for a single logratio or multivariate normal for a set of ratios. Such a possibility can have many benefits for statistical summaries and inference. This brings us directly to the topic of the lognormal distribution and its multivariate extensions, which is the subject of this chapter. The case of non-normal logratios will also be treated. This chapter can be skipped during a first reading of the book, or skipped entirely if the reader is interested only in data analytical aspects.

4.1 Lognormal distribution

The *normal distribution* is the most fundamental statistical probability distribution, thanks to the central limit theorem, which expresses that the mean of a sample of observations from almost any underlying statistical distribution becomes normally distributed for large samples. The *lognormal distribution* is the distribution of a positive-valued variable whose log-transformation is normally distributed. That is, putting this the other way round, if the random variable Y is normally distributed, then the exponential of Y, $X = \exp(Y) = e^Y$ is lognormally distributed, since $Y = \log(X)$ is normal.

Data following a normal distribution can take any real value, positive or negative, and are assumed to be on an additive (or interval) scale, where a comparison between two values is made by forming the difference. On the other hand, data that follow a lognormal distribution are strictly positive and are assumed to be on a multiplicative (or ratio) scale, where a comparison between two values is made as a ratio, i.e. as a percentagewise difference. For example, if X_1 and X_2 are lognormally distributed, then their logarithms $\log(X_1)$ and $\log(X_2)$ are observations from a normal distribution, with an implicit additive scale, so that their difference is measured by $\log(X_2) - \log(X_1) = c$, i.e. $\log(X_2) = c + \log(X_1)$. Exponentiated, the two lognormal observations are related multiplicatively as $X_2 = e^c X_1$, i.e. $X_1/X_2 = e^c$.

As an illustration, consider the ratio *Carbohydrate/Fat* (see Table 2.1). Fig. 4.1(a) shows the normal quantile plot for the 10 ratio values, obtained as follows. The data

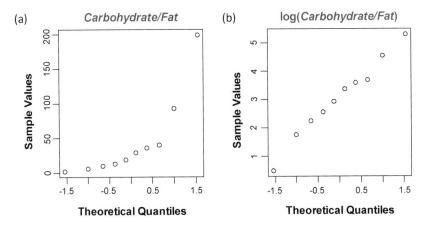

Fig. 4.1 Normal quantile plots to diagnose normality for (a) original ratio of *Carbohydrate/Fat*, and (b) log-transformed ratio, i.e. logratio, using R function qqnorm.

set is ordered from smallest to largest, this defines the vertical axis. The ranks 1 to n (here $n = 10$) are transformed to probabilities $p_i = (i - 0.5)/n$, $i = 1, 2, \ldots, n$ $= 0.05, 0.15, 0.25, \ldots, 0.95$. Then the standard normal distribution values corresponding to these cumulative probabilities, i.e. normal quantiles, are computed, giving the horizontal axis. If the data were from a normal distribution, the points would follow a straight line. So if they approximately follow a straight line, which is easy to judge, then the data are approximately normal. The convex shape of the plot in Fig. 4.1(a), however, indicates strong right skewness, which is typical for positive-valued data. Fig. 4.1(b) shows the plot for the corresponding logratios, and now the values fall more or less on a straight line, hence close to normal.

While the method of estimating the mean μ and variance σ^2 of the normal distribution is well-known, it is not so well-known for the lognormal distribution. Estimating the mean of a lognormally distributed variable is not simply a matter of computing the mean of the log-transformed values, and then back-transforming this mean with the exponential function — this would give an estimate that is biased downwards. The estimates of the mean and variance of the logratio values are:

$$\hat{\mu} = 3.043 \qquad \hat{\sigma}^2 = 1.912$$

The back-transformed value of the estimated mean of the ratios is $e^{\hat{\mu}} = e^{3.043} = 20.97$, but this severely underestimates the mean of the lognormally distributed ratios, which is correctly estimated as:

$$e^{\hat{\mu} + \hat{\sigma}^2/2} = 54.5$$

Hence, the back-transformed value of $e^{\hat{\mu}} = 20.97$ is adjusted upwards by multiplying by $e^{\hat{\sigma}^2/2} = 2.60$.

Since $\hat{\mu}$ estimates the median of the normally distributed logratios, the back-transformed value $e^{\hat{\mu}} = 20.97$ estimates the median of the lognormally distributed ratios — this model-based estimate can be compared with the sample median, 23.72, which splits the data into two halves. The *reference range* of a ratio includes a percentage, usually 95%, of the data. This range can be estimated using the estimated median and the 0.025 and 0.975 quantiles of the corresponding lognormal distribution as $[e^{\hat{\mu}}/1.96e^{\hat{\sigma}}, e^{\hat{\mu}} \times 1.96e^{\hat{\sigma}}] = [2.7, 163.8]$ (notice the multiply-and-divide operation for the range limits of this lognormal variable rather than the usual add-and-subtract operation for normal variables). This model-based interval can be compared with the sample estimates of the 0.025 and 0.975 quantiles of the data: $[2.6, 175.5]$, using R function `quantile`.

4.2 Logit function

Before considering how to generalize the idea of the lognormal to many ratios, it is necessary to dwell a moment on the *logit* transformation, synonymous with the concept of log-odds introduced in Sect. 1.5 and well-known in the area of *logistic regression* (see pages 36–37). The logit function takes this form: $y = \log[p/(1-p)]$, where p is a probability in the case of a log-odds, but in the compositional data context can be any compositional value between 0 and 1. Solving for p, the inverse function (often referred to as the logistic function) is $p = e^y/(1+e^y)$. For example, for a value $p = 0.8$, the logit transformation is $y = \log(0.8/0.2) = \log(4) = 1.386$. Using the inverse function, $e^{1.386}/(1+e^{1.386}) = 4/(1+4) = 0.8 = p$.

For a three-part composition $[x_1 \ x_2 \ x_3]$, like the one in Chap. 3, an additive logratio transformation can be defined using the last part x_3, for example, as the part in the denominator: $y_1 = \log(x_1/x_3)$ and $y_2 = \log(x_2/x_3)$. Then the inverse function to go back to the parts x_1, x_2 and x_3 is a generalization of the logistic function: $x_1 = e^{y_1}/(1+e^{y_1}+e^{y_2})$, $x_2 = e^{y_1}/(1+e^{y_1}+e^{y_2})$, $x_3 = 1 - x_1 - x_2$.

In general for a J-part composition, again using the last part in the denominator, as in (3.1), the ALRs and their inverse transformations are:

$$y_j = \log(x_j/x_J), \ j = 1, \ldots, J-1$$

$$x_j = e^{y_j}/(1 + \sum_{j=1}^{J-1} e^{y_j}), \ j = 1, \ldots, J-1 \quad x_J = 1 - \sum_{j=1}^{J-1} x_j \quad (4.1)$$

4.3 Additive logistic normal distribution

One of the possible multivariate distributions on a composition is thus based on their logratios being multivariate normal. For ALRs, the transformation back to the composition is performed using (4.1). The *additive logistic normal distribution* is thus the distribution of the parts if their $J - 1$ ALRs are multivariate normal. The parameters of the multivariate normal distribution are the mean vector $\boldsymbol{\mu}$ and the covariance matrix $\boldsymbol{\Sigma}$.

For the three-part data in Table 2.1, and using the ALRs $y_1 = \log(x_1/x_3) = \log(Protein/Fat)$ and $y_2 = \log(x_2/x_3) = \log(Carbohydrate/Fat)$, the estimates of these parameters are:

$$\hat{\boldsymbol{\mu}} = \begin{bmatrix} 1.846 \\ 3.043 \end{bmatrix} \quad \hat{\boldsymbol{\Sigma}} = \begin{bmatrix} 0.532 & 0.694 \\ 0.694 & 1.911 \end{bmatrix} \quad (4.2)$$

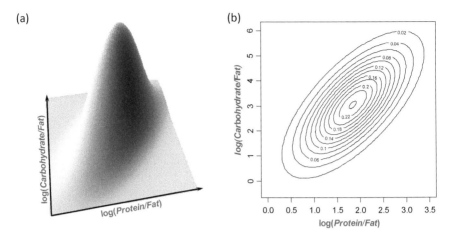

Fig. 4.2 Plot of bivariate normal distribution as (a) a perspective plot, and (b) a contour plot.

This normal distribution has a probability density function that is bell-shaped in three dimensions (Fig. 4.2(a)). The two dimensions for the two logratios define the two coordinates axes "on the ground", as it were, and the third dimension defines the height of the density function. The three-dimensional surface can be visualized in a perspective plot or a contour plot. The correlation between the two logratios is equal to 0.688, hence the elliptical contours in the contour plot of Fig. 4.2(b), with the major axes of the ellipses having positive slope.

4.4 Logratio variances and covariances

All the variances of the logratios can be collected in a square matrix called the *logratio variance matrix* $\mathbf{T}$, with elements $\tau_{jj'}$:

$$\tau_{jj'} = \text{var}[\log(x_j/x_{j'})], \quad j, j' = 1, 2, \ldots, J. \tag{4.3}$$

For the three-part vegetable data, where $x_1 = $ *Protein*, $x_2 = $ *Carbohydrate* and $x_3 = $ *Fat*, the estimated logratio variance matrix is:

$$\hat{\mathbf{T}} = \begin{bmatrix} 0 & 1.055 & 0.532 \\ 1.055 & 0 & 1.911 \\ 0.532 & 1.911 & 0 \end{bmatrix} \tag{4.4}$$

The values in either the upper or lower triangle of this matrix quantify the total variance in the compositional data set. A convenient definition of the total logratio variance is to sum these variances of unique pairwise logratios and divide by J^2 ($3^2 = 9$ in this example). However, for this purpose the maximum likelihood estimates of the variances will be preferred: that is, in the variance calculation the sum of squared deviations is divided by the sample size I, not by $I-1$ (hence, these estimates of the variance are slightly biased downwards). Thus, for $I = 10$ in this example, the usual sample variances in (4.4) have to be adjusted down by $9/10 = 0.9$, so the total logratio variance would be $0.9 \times (1.055 + 0.532 + 1.911)/9 = 0.350$. The subject of measuring logratio variance will be explained in more detail in Sect. 5.3.

The variances and covariances of all the unique pairwise logratios define the *logratio covariance structure* of the compositional data set, a $\frac{1}{2}J(J-1) \times \frac{1}{2}J(J-1)$ matrix Σ^* with elements

$$\sigma^*_{jj',kk'} = \text{cov}\left(\log(x_j/x_{j'}), \log(x_k/x_{k'})\right), \quad j < j', k < k' = 1, 2, \ldots, J. \quad (4.5)$$

For example, for the three-part vegetable data, this gives a 3×3 matrix $\hat{\Sigma}^*$ of estimated covariances of the logratios $\log(x_1/x_2)$, $\log(x_1/x_3)$ and $\log(x_2/x_3)$:

$$\hat{\Sigma}^* = \begin{array}{c} \\ 12 \\ 13 \\ 23 \end{array} \begin{array}{ccc} 12 & 13 & 23 \\ \left[\begin{array}{ccc} 1.055 & -0.162 & -1.217 \\ -0.162 & 0.532 & 0.694 \\ -1.217 & 0.694 & 1.911 \end{array}\right] \end{array} \quad (4.6)$$

The diagonal values are the variances (cf. the variances in (4.4)) and the covariance in row 12 and column 13, for example, is between the logarithms of *Protein/Carbohydrate* and *Protein/Fat*, equal to -0.162. The matrix $\hat{\Sigma}^*$ is symmetric, so the unique values are in the upper (or lower) triangle as well as the diagonal, giving $\frac{1}{2}J(J-1) + J = \frac{1}{2}J(J+1)$ values, i.e. 6 unique values in this $J = 3$ case.

A unique and surprising result, which is particular to compositional data analysis due to the dependencies between logratios, is that all the covariances in (4.5) can be determined from the logratio variances in (4.3) through the relationship:

$$\sigma^*_{jj',kk'} = \frac{1}{2}\left(\tau_{jk'} + \tau_{j'k} - \tau_{jk} - \tau_{j'k'}\right) \quad (4.7)$$

For example, the estimated covariance $\sigma^*_{12,13} = -0.162$ mentioned above is (see (4.4)):

$$\hat{\sigma}^*_{12,13} = \frac{1}{2}\left(\tau_{13} + \tau_{21} - \tau_{11} - \tau_{23}\right)$$
$$= \frac{1}{2}\left(0.532 + 1.055 - 0 - 1.911\right)$$
$$= -0.162.$$

Finally, to return to the covariance matrix $\Sigma = [\sigma_{jj'}]$ of the additive logratios (see Sect. 4.3, with covariance estimates in (4.2)), this $(J-1) \times (J-1)$ matrix has $\frac{1}{2}J(J-1)$ unique values, made up of $\frac{1}{2}(J-1)(J-2)$ covariances and $J-1$ variances. These values also determine all the unique elements of the covariance structure. For example, the covariances in (4.2) for the vegetable data form the bottom right hand 2×2 submatrix of (4.6). The remaining two covariances and variance in the first row (or first column) of (4.6) can be obtained from Σ, using the general formula:

$$\sigma^*_{jj',kk'} = \sigma_{jk} + \sigma_{j'k'} - \sigma_{jk'} - \sigma_{j'k} \tag{4.8}$$

For example, the variance of $\log(x_1/x_2)$ in the upper left hand corner of (4.6), reduces to the well-known formula for the covariance of the sum of two variables: $\mathrm{var}(A+B) = \mathrm{var}(A) + \mathrm{var}(B) - 2\mathrm{cov}(A,B)$, where $A = \log(x_1/x_3)$ and $B = \log(x_2/x_3)$ so that $\mathrm{var}(A+B) = \mathrm{var}(\log(x_1/x_2))$:

$$\begin{aligned}
\hat{\sigma}^*_{12,12} &= \hat{\sigma}_{11} + \hat{\sigma}_{22} - \hat{\sigma}_{12} - \hat{\sigma}_{21} \\
&= 0.532 + 1.911 - 0.694 - 0.694 \\
&= 1.055
\end{aligned}$$

The next value (-0.162) in (4.6), the covariance between $\log(x_1/x_2)$ and $\log(x_1/x_3)$ is:

$$\begin{aligned}
\hat{\sigma}^*_{12,13} &= \hat{\sigma}_{11} + \hat{\sigma}_{23} - \hat{\sigma}_{13} - \hat{\sigma}_{21} \\
&= 0.532 + 0 - 0 - 0.694 \\
&= -0.162
\end{aligned}$$

Notice that $\hat{\sigma}_{23}$ and $\hat{\sigma}_{13}$ are zero because the additive logratios are with respect to part 3, so $\hat{\sigma}_{23} = \mathrm{cov}(\log(x_2/x_3), \log(x_3/x_3)) = 0$, for example. Similarly, the last value in the first row of (4.6) is $\hat{\sigma}_{12} - \hat{\sigma}_{22} = 0.694 - 1.911 = -1.217$.

In the simple three-part compositional example we have seen how a set of two additive logratios has a covariance matrix from which the complete covariance structure can be determined. In general, a set of $J-1$ additive logratios has a covariance matrix that determines the complete covariance structure of a compositional data set. This set depends on which part is chosen as the denominator, and the obvious question arises as to what happens if a different part is chosen. Clearly, the covariance matrix will change; for example, if the first part (*Protein*) had been chosen in the vegetables example, the covariance matrix would have been the upper left hand 2×2 matrix of (4.6), and this would generate a different bivariate normal distribution, this time with negative correlation. The important question is: does it make any difference to what we do later with the covariance matrix? Do results such as assessing group differences, or identifying important features in the data set, change if a different set of additive logratios is chosen? In other words, it is clear that the additive logratios themselves are not permutation invariant, but is this of any eventual consequence to our conclusions? This question will be answered in later sections and chapters, where relevant.

4.5 Testing for multivariate normality

If the additive logratios are multivariate normal, then there are several procedures in classical multivariate analysis that can be validly used, for example confirmatory factor analysis, discriminant analysis and multivariate analysis of variance. Only one approach will be considered here, the multivariate generalization of the *Shapiro-Wilks test*.

In Fig. 4.1 normal quantile plots were shown, which compared the ordered data to the corresponding quantiles of a normal distribution. The closer the points were to a straight line, the closer they were to being normally distributed. The Shapiro-Wilks test is one way of testing the "straightness" of this relationship. It is not a question of testing a regular correlation coefficient, because the points are not a set of observed bivariate data — here it is one set plotted against a hypothesized set of theoretical values. But the statistic W used in the test can be thought of as a correlation-type measure, lying between 0 and 1: the higher W is, the closer to normality are the data. Luckily, the distribution of W under the hypothesis of normality is known, and so p-values can be computed.

For the ratio *Carbohydrate/Fat* and its logarithm in Fig. 4.1, the results of the test, using the R function `shapiro.test`, are $W = 0.702$ ($p = 0.0009$) and $W = 0.991$ ($p = 0.99$) respectively, showing how far from normality the ratio is, but how close to normality the logratio is. The other logratios are also judged to be normal, in the sense that normality is not rejected by the hypothesis test. For log(*Protein/Fat*), $W = 0.952$ ($p = 0.70$), and for log(*Carbohydrate/Protein*), $W = 0.895$ ($p = 0.19$).

Since all the pairwise logratios are judged to be close to normal, this gives some credence to the set of ALRs being multivariate normal, but this is not a sufficient condition. The multivariate generalization of the Shapiro-Wilks test should rather be used, and there are several variants and R packages available. Using the package `mvtnorm`, for example, the test statistic has the same value, no matter which set of ALRs is tested: $W = 0.980$ ($p = 0.97$). Other packages with different procedures, such as `MVN` and `royston`, have test statistics that give different values for each set of ALRs, but none of these tests rejects multivariate normality.

4.6 When logratios are not normal

Having logratios that are normally distributed is convenient because there are many so-called "parametric" statistical procedures that can be used, for testing hypotheses and setting confidence intervals, for example. But non-normality of data is quite common in statistics, so various approaches have been developed to cope with it. For non-normal univariate logratios, non-parametric methods are available, as well as resampling methods such as permutation testing.

For example, in Fig. 3.2 there was a split between two groups of vegetables, {Potatoes, Onions, Peas, Asparagus, Spinach, Broccoli} and {Carrots, Corn, Mushrooms, Beans}. So, supposing that it was interesting right from the start to test this difference, the fact that the logratios are deemed to be normally distributed would permit the use of the classical two-group t-test between these two groups. The t-tests for the three logratios turn out as follows:

$$t_{Prot/Carb} = 0.557 \quad t_{Prot/Fat} = 5.62 \quad t_{Carb/Fat} = 0.954$$
$$(p = 0.60) \qquad (p = 0.0008) \qquad (p = 0.38)$$

Clearly, it is the *Protein/Fat* logratio that distinguishes the two groups of vegetables and the difference is highly significant (see Figs 2.3 and 2.7 where this logratio lines up with the group difference).

If the data were not normal, there are several approaches, but one of the best is to use distribution-free permutation testing, available in the R package `coin`. Performing permutation tests for contrasting the same two groups of vegetables gives the following results, similar to the t-tests in their conclusions:

$$Z_{Prot/Carb} = 0.588 \quad Z_{Prot/Fat} = 2.67 \quad Z_{Carb/Fat} = 0.971$$
$$(p = 0.57) \qquad (p = 0.005) \qquad (p = 0.35)$$

To test multivariate differences between the two groups, that is comparing all logratios simultaneously, a multivariate analysis of variance (MANOVA) can be used when the data are multivariate normal. Alternatively, a multivariate permutation test can be used for non-normal data. For the MANOVA test, a set of linearly independent logratios has to be used, so for the vegetables data, any set of ALRs will do, and it does not matter which is chosen, the result will be the same. Using the R function `manova`, the MANOVA test gives a p-value for group difference as $p = 0.00003$, a highly significant result.

Using a multivariate distance between the two group means, which serves as a measure of difference between them, a distribution-free permutation test can be conducted as follows. The distance between the groups is computed and saved — this is the "original" test statistic. Then the group categories, 6 group 1s and 4 group 2s in this case, are randomly assigned to the 10 vegetables, under the hypothesis of no group difference, and the inter-group distance recomputed. This permutation of the group labels and recomputation of the inter-group distance is repeated a large number of times, 999 times in this case, and then the number of distances greater than or equal to the original one is counted. It turns out that there are only 5 such distances out of the total of 1000 (including the original one), so this gives an estimated p-value of $p = 5/1000 = 0.005$, indicating a significant difference.

Summary: Properties and distributions of logratios

- A single set of logratios can be assessed to be approximately normal or not by plotting their values against the normal quantiles. A statistical test, the Shapiro-Wilks test, is one of the ways of formally testing for normality.
- If a logratio is normally distributed, then the ratio itself is lognormally distributed.
- A set of logratios can be approximately multivariate normal — this can be verified using a generalization of the Shapiro-Wilks test.
- The additive logistic normal distribution is the distribution of a set of independent ratios, where their additive logratios are normally distributed.
- The covariance matrix of all the $\frac{1}{2}J(J-1)$ pairwise logratios can be determined knowing just their $\frac{1}{2}J(J-1)$ variances, or the covariance matrix of $J-1$ independent logratios, for example a set of ALRs.
- When logratios are not normally distributed, nonparametric methods are available as well as permutation testing.

Chapter 5
Regression models involving compositional data

In statistical modelling, compositions can be used as predictors (independent variables, or explanatory variables) or as responses (dependent variables). As stressed throughout this book, logratio transformations of the compositions are essential, and for J-part compositions a set of $J - 1$ additive logratios, or a set of $J - 1$ independent logratios is sufficient to represent all the variance of the compositional data set. In some circumstances it may be simplifying to reduce the number of logratios by some type of variable selection process. This is always possible because logratios will always be correlated with one another, so a smaller subset of them might well be adequately "close" enough to the original set for all practical purposes.

5.1 Visualizing ratios as a graph

Before getting onto the subject of this chapter, here is a short section about visualizing a set of ratios, especially for checking that they are independent of one another. Ratios can be depicted in a *directed graph*, where the parts are connected by arrows indicating the respective ratio and pointing towards the numerator. An example is given in Fig. 5.1(a) of two ratios A/C and A/B. In Fig. 5.1(b) four ratios are shown, but the addition of the ratio D/B to the others would create a *cycle* in the graph, and make the set of ratios in the cycle dependent on one another. For example, $D/B = A/B \times D/A$, where the path from B to D can be traced from B to

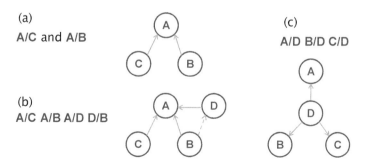

Fig. 5.1 (a) Two logratios (arrow points to numerator in each case); (b) The logratio D/B indicated by a dashed arrow creates a cycle; (c) ALRs of a four-part composition {A,B,C,D}, where D serves as the denominator.

A and then from A to D — notice that the arrow in Fig 5.1(b) points from D to A to represent A/D, so the arrow in the opposite direction inverts the ratio to be D/A. When logarithmically transformed the relationship becomes linear: $\log(D) - \log(B)$ = [$\log(A) - \log(B)$] + [$\log(D) - \log(A)$]. It follows that, for a set of logratios to be useful explanatory variables in a regression, they should not form a cycle in their graph representation. Finally, ALRs such as Fig. 5.1(c) form a star pattern with the denominator part in the centre.

5.2 Using simple logratios as predictors

A compositional data set can serve as predictors of one or more response variables, or it can constitute the responses themselves, predicted by additional variables. In this section, the first case is concerned, and is applied to two data sets, called TimeBudget and FishMorphology.

Data set TimeBudget

This data set, shown in Table 5.1, consists of the time in hours spent on average by males and females in 16 countries, over a period of 24 hours — each row of the table thus sums to 24. Now suppose it is of interest to study the varying amounts of sleep as a response variable, by relating it to the activities during the day. There is no point in modelling the sleep hours as a function of the other hours directly, since the relationships will generally be negative: if day activities increase in time, sleep hours have to go down, since the total number of hours in a day is fixed. What would be more interesting is to relate the sleep hours to the ratios of day activities; for example, does the amount of sleep depend on the ratio of work to free time? There are five day activities and thus 10 possible ratios. Using each logratio in turn as a predictor, it turns out that the logarithm of the ratio *meals/travel* is the best linear predictor of *sleep*, the estimated regression model being:

$$\text{mean}(\textit{sleep}) = 7.99 + 0.375 \times \log(\textit{meals/travel})$$

with the p-value for the regression being $p = 0.025$ ($R^2 = 0.157$). The demographic groups with the highest and lowest ratios of *meals/travel* are, respectively, French females (FRf, value of ratio = 2.73) and British males (UKm, ratio = 1.20). From the regression their predicted average hours of sleep (with observed values in parentheses) are, respectively, 8.37 (8.63) and 8.06 (8.18). The model suggests that the more time spent eating compared to travelling, the higher the average sleep time.

Having chosen this ratio, the process can be continued like a stepwise regression analysis, selecting the next best predictor, but there are no other ratios that add significantly to the above single-predictor regression model.

Data set FishMorphology

The second data set consists of 26 morphometric measurements, in millimeters, on a sample of 75 fish of the species Arctic charr (*Salvelinus alpinus*), shown in Fig. 5.2.

Table 5.1 Data set `TimeBudget`: average hours of daily activity and sleep in 16 countries, split by males and females.

	Average Daily Hours					
	work	*domestic*	*travel*	*sleep*	*meals*	*free*
TRm	6.22	0.72	1.75	8.13	2.70	4.48
BEm	5.05	2.25	1.72	8.02	2.58	4.38
DEm	5.08	1.87	1.52	8.00	2.35	5.18
FRm	5.73	1.88	1.17	8.40	2.97	3.85
HUm	5.58	2.22	1.20	8.38	2.46	4.16
FIm	5.53	1.98	1.28	8.20	1.92	5.08
SEm	5.28	2.38	1.53	7.87	2.08	4.85
UKm	5.70	1.90	1.60	8.18	1.92	4.70
NOm	4.93	2.20	1.38	7.88	1.97	5.63
EEm	5.00	2.33	1.33	8.37	2.18	4.78
ITm	6.22	1.17	1.67	7.97	2.87	4.12
ESm	6.18	1.33	1.38	8.25	2.52	4.33
LTm	6.52	1.65	1.28	8.13	2.38	4.03
LVm	6.68	1.43	1.52	8.27	2.13	3.97
PLm	6.17	1.88	1.25	7.98	2.23	4.48
SIm	5.33	2.40	1.23	8.10	2.12	4.82
TRf	4.41	4.05	1.15	8.11	2.57	3.72
BEf	3.88	3.87	1.50	8.27	2.60	3.88
DEf	3.87	3.18	1.45	8.18	2.52	4.80
FRf	4.53	3.67	1.08	8.63	2.95	3.13
HUf	4.72	3.90	1.03	8.30	2.35	3.70
FIf	4.33	3.35	1.27	8.37	2.03	4.65
SEf	4.08	3.53	1.47	8.08	2.38	4.45
UKf	4.10	3.47	1.55	8.42	2.12	4.35
NOf	3.77	3.43	1.28	8.12	2.03	5.37
EEf	4.22	4.07	1.25	8.38	2.10	3.98
ITf	4.65	3.85	1.47	8.00	2.73	3.30
ESf	4.95	3.48	1.37	8.18	2.47	3.55
LTf	5.92	3.40	1.12	8.22	2.27	3.08
LVf	5.77	3.13	1.43	8.35	2.10	3.22
PLf	4.77	3.97	1.17	8.14	2.23	3.72
SIf	4.38	4.40	1.15	8.20	2.03	3.83

TR	Turkey
BE	Belgium
DE	Germany
FR	France
HU	Hungary
FI	Finland
SE	Sweden
UK	United Kingdom
NO	Norway
EE	Estonia
IT	Italy
ES	Spain
LT	Lithuania
LV	Latvia
PL	Poland
SI	Slovenia
m	male
f	female

Additionally the mass of each fish is recorded, ranging from 62 to 178 grams, as well as its sex (F = female; M = male) and habitat (L = littoral, close to the shore; P = pelagic, in deeper water far from the shore). A small part of the data set is shown in Table 5.2. Ratios of pairs of these measurements quantify the shape of the fish, since the size of each fish is eliminated. However, these shape parameters can still

Table 5.2 Part of the data set `FishMorphology`, *showing the first six rows out of 75, and the first 12 morphometric measurements (mm) out of 26, the mass (g), sex and habitat.*

FishID	Hw	Bg	Bd	Bcw	Jw	Jl	Bp	Bac	Bch	Fc	Fdw	Faw	⋯	Mass	Sex	Habitat
19	20.9	24.6	24.6	8.9	9.7	28.3	83.3	33.3	14.9	34.2	25.9	21.0	⋯	125	F	P
23	19.5	24.3	25.0	8.4	8.6	26.6	75.6	32.9	15.3	28.9	25.2	19.0	⋯	110	M	L
24	20.8	26.3	28.7	11.0	10.8	27.6	81.6	34.2	16.5	29.7	26.3	17.8	⋯	140	M	L
25	23.1	28.5	30.9	10.3	11.2	29.6	85.4	35.0	18.1	33.5	25.3	20.9	⋯	178	M	L
27	21.4	26.5	28.4	10.2	11.4	31.0	82.5	32.4	17.6	32.8	29.5	21.6	⋯	154	F	L
28	22.9	28.0	30.8	10.4	11.6	28.8	77.7	32.6	18.4	30.6	26.6	20.1	⋯	146	M	L
⋮	⋮	⋮	⋮	⋮	⋮	⋮	⋮	⋮	⋮	⋮	⋮	⋮		⋮	⋮	⋮

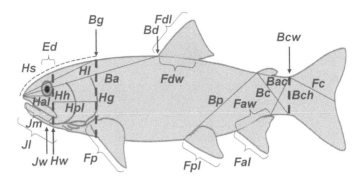

Fig. 5.2 Morphometric measurements on the *Arctic charr*.

be related to the mass of the fish or to the categories of sex and habitat. Notice that these data are not strictly compositional but they can be treated in the same way as compositional data, if the 26 morphometric measurements are closed by expressing them as proportions of their respective sums. Computing ratios on the original data is the same as computing them on the closed data.

The stepwise process of selecting logratios proceeds in two steps: Table 5.3 shows the ratios selected, estimated regression coefficients and respective p-values in the final model where the response is the logarithm of the mass. It is well-known that the mass (M) of a fish of a particular species can be accurately estimated by knowing its length (L) and width (W), which usually results in a power model for that species of the form $M \approx cL^{\alpha}W^{\beta}$, where $\alpha + \beta \approx 3$ for some constant c (this is a linear model when log-transformed). In the present application, however, we are investigating whether the mass can be explained by the shape of the fish, i.e. the relative values of the morphometric measurements, without knowledge of any absolute values.

Table 5.3 Terms in multiple regression model of mean log(mass) as a function of four logratios of morphometric measurements (see Fig. 5.2). The variance explained is 36.5%. R function `lm` is used.

Term	Coefficient	p-value
Intercept	6.09	($p < 10^{-15}$)
log(*Bcw/Fdw*)	1.90	($p < 10^{-7}$)
log(*Faw/Bc*)	0.618	($p = 0.006$)

We can similarly investigate whether the shape of the fish can predict its sex, using stepwise logistic regression, for example (available in the `glm` function in R). It turns out that the best logratio for predicting sex is *Faw/Fdl* and no other logratio adds significantly more to the model. The logistic regression models the log-odds of the fish being male (see discussion about log-odds on pages 7 and 27). If p is the probability of being male, the estimated model is:

$$\log\left(\frac{p}{1-p}\right) = -3.69 - 12.42 \times \log\left(\frac{Faw}{Fdl}\right)$$

with the p-value for the logratio predictor of $p = 0.0004$. Using the cutoff probability of 0.5, which is equivalent to the log-odds of 0, the prediction of sex is successful 26 out of 37 times for the females and 28 out of 38 times for the males, i.e. $11 + 10 = 21$ incorrect predictions.

An alternative is to use classification tree methodology (`rpart` function in R). Since the log-transformation is monotonically increasing, the cutoffs will be equivalent in the partitioning of the variables, transformed or untransformed, so the ratios can be left in their original scale. The estimated tree is given in Fig. 5.3. The tree shows the same ratio *Faw/Fdl* predicting females for higher values of the ratio, agreeing with the logistic regression model on the previous page. But for lower values of this ratio, two other ratios play a role, resulting in more success in estimating the females (36 out of 37 correctly predicted) but slightly less in predicting males (27 out of 38 males correctly predicted), i.e. $1 + 11 = 12$ false predictions in total. Of course, there is the issue of cross-validation for the classification tree approach as well as the logistic regression, which we do not enter into here, given the small sample.

Fig. 5.3 Classification tree predicting sex of Arctic charr. The inequality at each node indicates the condition that sends part of the sample to the left. The terminal nodes are the female/male counts, with predicted sex in parentheses. For example, the first split, for *Faw/Fdl* > 0.758 takes 29 fish to the left, of which 24 are females, hence they are predicted females. So the 5 males at this terminal node are incorrectly predicted as females.

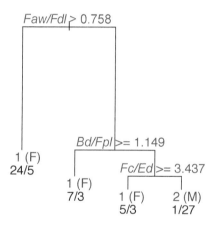

5.3 Compositions as responses – total logratio variance

The situation is now turned around, and the compositions are considered the responses to some other predictors, which is perhaps the more common situation in practice. Again, a process of ratio selection could be adopted, selecting the logratio responses that are well explained by the predictors, but here it is usual to take the full set of logratios as the responses, at least in a first analysis.

In regression analysis the single response, or dependent variable, has just one variance, and this variance is explained by the predictors, or independent variables. When many variables are regarded as a set of responses to be explained, each with its own variance, a definition is needed of the total variance, and this definition should combine the individual variances of the responses in an equitable way. There are several ways to define the total variance, two of which are considered here, firstly based on all the pairwise logratios and secondly based on the centred logratios. The distinction will also be made between unweighted and weighted measures of total variance.

Total variance of the logratios

Suppose the compositional data set is $\mathbf{X}(I \times J)$ and the variance of the (j, j')-th logratio is denoted by $\text{var}_{jj'}$. Here it will be preferable, as already mentioned in Sect. 4.4, to define the variance as the average sum-of-squared deviations of the logratio values from their mean — that is, dividing the sum-of-squares by I and not by $I - 1$. Hence, compared to the usual unbiased variance denoted by $\tau_{jj'}$ in (4.3), $\text{var}_{jj'} = \tau_{jj'} \times (I - 1)/I$ — notice that the difference between $\text{var}_{jj'}$ and $\tau_{jj'}$ gets less and less as I increases, and is trivial for large I. Then the total unweighted variance of the compositional data set is defined as $(1/J^2) \sum_{j<j'} \text{var}_{jj'}$, where $\sum_{j<j'}$ denotes the summation over all unique pairs of parts.

Notice that the summation involves $\frac{1}{2}J(J-1)$ logratio variances, but the variances are not averaged but rather divided by J^2. This is because $1/J$ is regarded as the weight assigned to each of the J parts, and each logratio variance is multiplied by the product of the weights of its two parts which is $1/J^2$. This definition can be generalized to the *weighted* form of the total variance, where each part receives a different weight according to a prescribed set of positive values $c_j, j = 1, \ldots, J$, where $\sum_j c_j = 1$. In this case the definition of the total weighted logratio variance is:

$$\text{total weighted variance} = \sum_{j<j'} c_j c_{j'} \text{var}_{jj'} \qquad (5.1)$$

with the unweighted case being the special case when $c_j = 1/J$ for all j.

Total variance of the centred logratios

Computing the total variance based on all logratios requires the computation of $\frac{1}{2}J(J-1)$ variances. Exactly the same result can be obtained computing the J variances based on the centred logratios. Suppose the centred logratios (CLRs), i.e. the columns of the row-centred log-transformed compositional data matrix, have variances denoted by $\text{var}_j, j = 1, \ldots, J$, with a single subscript, where again variances are computed by dividing the sum-of-squared deviations from the mean of each CLR by I, not $I - 1$. Then the total unweighted variance is equal to $(1/J) \sum_j \text{var}_j$. For the more general weighted version, the CLRs have first to be recomputed with respect to weighted row means of the log-transformed table (see (A.5) in Appendix A), before computing their variances var_j:

$$\text{total weighted variance} = \sum_j c_j \text{var}_j \qquad (5.2)$$

The unweighted case is this time not simply the special case of (5.2) when $c_j = 1/J$ for all j, but requires CLRs centred by regular arithmetic averages, not weighted, before computing the variances. Using the CLRs saves computational effort, since only J variances are computed, not $\frac{1}{2}J(J-1)$.

5.4 Redundancy analysis

An approach to modelling a set of responses in terms of one or more explanatory variables is called *redundancy analysis* (RDA, available in the R package `vegan`). RDA, also called reduced-rank regression or principal component analysis of instrumental variables, is principally used for dimension reduction, to be dealt with in Chap. 6. At present, however, the only use we will make of RDA is to quantify how much variance is explained and to test which of the explanatory variables account for significant variance in the compositional responses.

So suppose the objective of the `TimeBudget` study is turned around: that is, it is being investigated how strongly the amount of sleep explains the subcomposition formed by the other five daily activities. The response matrix can be the 10 logratios of the five-part subcomposition or the five CLRs themselves, the result will be the same in terms of percentage of variance explained. The total (unweighted) logratio variance of the subcomposition is equal to 0.05184, of which sleep explains 0.00403 (7.78%). The `vegan` package also contains a permutation for testing whether the explained variance is significant or not, and the estimated p-value is $p = 0.09$, hence this amount of variance explained is not significant at the 0.05 level.

Considering now the `FishMorphology` data set, the shapes of the fish are coded by the morphometric measurements considered relative to their total. Explanatory variables available for explaining logratio variance of shape are *Mass*, *Sex* as well as the additional binary variable *Habitat*, whether the fish is littoral or pelagic. In choosing between unweighted or weighted total variance, weighted variance is now chosen since the lower-valued measurements have higher logratios compared to the higher-valued ones. The weighting will balance out the contributions of low- and high-valued measurements on the fish to the total logratio variance, which would otherwise be dominated by the smaller measurements. This is a recurring issue in compositional data analysis and will be discussed more fully in Chapter 6.

With three explanatory variables available, a stepwise approach can again be used, so each variable is used in a RDA to see how much variance it explains. The total weighted logratio variance is equal to 0.001968, much less than that of the five-part subcomposition in the previous `TimeBudget` example. This is because the fish are very similar to one another in shape, all being of the same species. The three explanatory variables account for the following percentages of variance, along with their respective p-values, in separate RDAs:

log(*Mass*): 3.6% ($p = 0.004$)   *Sex*: 2.2% ($p = 0.08$)   *Habitat*: 2.8% ($p = 0.02$)

The percentages are small, but two of them are significant at the 0.05 level, with log(mass) accounting for the most variance. So log(*Mass*) is fixed in the model and the other two are tried in turn as a second explanatory variable. The results, when added to log(*Mass*), are given as additional variance explained and associated p-values, again in separate RDAs:

Sex: 2.1% ($p = 0.08$)   *Habitat*: 2.8% ($p = 0.02$)

Sex is again not significant, but *Habitat* adds significant explained variance.

A further RDA with all three explanatory variables shows again that sex is not significant, so log(*Mass*) and *Habitat* are retained as explanatory variables, with a total of 6.6% explanation of the shape.

This resulting "model" is thus that the compositional data set is related linearly to log(mass) and habitat, but what are the "coefficients" of the relationship to allow us to understand what the relationship is? To answer this, each logratio of the compositional data set can be related to log(mass) and habitat in order to pick out the strongest relationships. Of all the $\frac{1}{2} \times 26 \times 25$ logratios of the morphometric measurements, those in Table 5.4 are the five most strongly explained by the two explanatory variables.

Response	Intercept	Coeff. log(*Mass*) (p-value)	Coeff. *Habitat (pelagic)* (p-value)	R^2 (%)
Ed/Hg	−0.722	−0.123 ($p < 0.0001$)	0.0509 ($p = 0.004$)	27.8
Hg/Bac	−0.190	0.059 ($p = 0.02$)	−0.0700 ($p < 0.0001$)	26.7
Ed/Bcw	0.928	−0.188 ($p < 0.0001$)	0.0203 ($p = 0.04$)	24.6
Ed/Jl	−0.526	−0.111 ($p < 0.0001$)	−0.0006 ($p = 0.97$)	22.6
Bcw/Hw	−1.300	0.107 ($p < 0.0001$)	0.0127 ($p = 0.40$)	21.6

Table 5.4 Terms in separate multiple regression models of mean logratios as responses to the explanatory variables log(*Mass*) and *Habitat*. The estimated coefficient for *Habitat* refers to the difference between *pelagic* and *littoral*.

Summary: Regression models involving compositional data

- In a regression context, compositional data can be considered as a set of possible explanatory variables or response variables.
- As explanatory variables, all pairwise logratios are candidates and the problem is one of variable selection, for example using a stepwise approach. The logratio that explains the maximum variance (or leads to minimum deviance in a generalized linear model) is chosen first and fixed. Then the second best logratio is chosen and so on, until no significant additional explanation of the response results.
- As response variables, the complete set of logratios is to be explained by additional explanatory variables. It is equivalent to use centred logratios as the response set.
- Redundancy analysis is a generalization of regression when there are multiple responses. A key result is the percentage of variance in the set of responses that is explained by the predictors (explanatory variables). Explanatory variables can also be entered in a stepwise manner.
- Having established a set of explanatory variables that significantly explains a compositional data set, individual logratio responses can be investigated to isolate those that are best explained.

Chapter 6
Dimension reduction using logratio analysis

Compositional data are necessarily multivariate and multivariate data always consist of intercorrelated variables. This means that the essential information in the data lies mostly in fewer dimensions than the original data and the object is then to identify these important variance-explaining dimensions. Logratio analysis is a variant of principal component analysis, applied not to the original compositional data, but rather to the complete set of logratios, or equivalently the set of centred logratios. The issue of how to weight the compositional parts is important in logratio analysis, since some parts can have higher relative error than others. A compositional data matrix regarded as a set of responses to a set of explanatory variables can be analysed by redundancy analysis to investigate compositional dimensions directly related to these predictors.

6.1 Weighted principal component analysis

Principal component analysis (PCA) is a well-known method for identifying a few dimensions that account for, or explain, a maximum part of the variance in a data set. PCA applies to observations on a set of continuous, interval-scale variables. PCA can be equivalently thought of as reducing the dimensionality of a covariance or correlation matrix, and so is not suitable for analysing the original compositional data because of the lack of subcompositional coherence mentioned in Chap. 1. PCA is rather applied to the matrix of logratios or, equivalently, the matrix of centred logratios. It is then known as *logratio analysis* (LRA) and has special properties due to the nature of the data structure.

Before proceeding to LRA, the general case of weighted PCA is considered, of which LRA is a special case. Weighted PCA is the weighted least-squares matrix approximation of a given data matrix, suitably transformed, by another matrix of lower rank, where rank is a synonym for dimensionality. The least-squares approximation involves weights for both the rows and columns, which can improve or diminish the role of chosen rows or columns to determine the low-rank approximation. In other words, a fixed matrix is being approximated by another one of lower rank, but the approximation will tend to be better for those rows or columns that have higher weights. This is useful if some variables have higher measurement error, in which case their weight in the analysis can be reduced.

Suppose the matrix to be approximated is $\mathbf{Y}$ ($I \times J$) and the row and column weights are in the vectors $\mathbf{r}$ ($I \times 1$) and $\mathbf{c}$ ($J \times 1$) respectively. The weights are all positive and usually sum to 1 in each case. The following optimization problem needs to be solved:

$$\underset{\hat{\mathbf{Y}} \text{ of rank } K}{\text{minimize}} \sum_i \sum_j r_i c_j (y_{ij} - \hat{y}_{ij})^2 \tag{6.1}$$

In other words, the solution $\hat{\mathbf{Y}}$ should come as close as possible to the given matrix $\mathbf{Y}$, but be of much lower rank K, i.e. K is much less than I or J, whichever is smaller. And more squared error $(y_{ij} - \hat{y}_{ij})^2$ can be tolerated for rows or columns with lower weights than those with higher weights.

The minimization in (6.1) can be solved easily using the singular value decomposition (SVD), one of the most useful results in matrix algebra, and available in the R function `svd`. More details are given in Appendix C, here it is simply stated that a matrix $\mathbf{S}$ is first computed with elements $s_{ij} = \sqrt{r_i c_j} y_{ij}$ (take r_i and c_j into the parentheses in (6.1)). Then the SVD of the matrix $\mathbf{S}$ is computed, which provides immediately a low-rank approximation $\hat{\mathbf{S}}$ of $\mathbf{S}$ of any desired dimensionality (usually 2 is chosen). Then, to get the approximation to the original $\mathbf{Y}$, simply divide the elements of the approximation to $\hat{\mathbf{S}}$ by $\sqrt{r_i c_j}$.

6.2 Logratio analysis

In LRA the matrix $\mathbf{Y}$ is either the matrix of logratios or the matrix of CLRs — in both cases the logratios are then centred columnwise, i.e. column means are subtracted out. Since the matrix of CLRs has considerably fewer columns than the matrix of logratios, it is the matrix of CLRs that is analysed, being mathematically equivalent to analysing all the logratios. This is the previously mentioned computational benefit of using CLRs — the results for the $\frac{1}{2}J(J-1)$ pairwise logratios can be obtained from those of the J CLRs. The weights $\mathbf{r}$ for the rows (samples) are usually equal, i.e. $r_i = 1/I$ for all rows $i = 1, \ldots, I$. Default weights $\mathbf{c}$ for the columns (compositional parts) are the means of the respective parts. Both $\mathbf{r}$ and $\mathbf{c}$ sum to 1.

Let us consider a new data set called `RomanCups`, the composition of oxides in 47 Roman glass cups discovered in eastern England. A small part of this 47×11 archaeometric data table is shown in Table 6.1, along with some column means.

Table 6.1 Part of the data set `RomanCups`, showing the first six rows out of 47, and five of the 11 mineral oxides in the composition. The rows sum to 100%. *Si* – silicon, *Al* – aluminium, *Fe* – iron, *P* – phosphorus, *Mn* – manganese, *Sb* – antimony.

CupID	Si	Al	Fe	$\cdots$	P	Mn	Sb
1	75.2	1.84	0.26	$\cdots$	0.04	0.01	0.36
2	72.4	1.80	0.28	$\cdots$	0.04	0.01	0.33
3	69.9	2.08	0.40	$\cdots$	0.06	0.03	0.44
4	70.2	2.23	0.41	$\cdots$	0.05	0.01	0.34
5	73.0	2.16	0.35	$\cdots$	0.05	0.01	0.37
6	74.2	2.02	0.33	$\cdots$	0.05	0.01	0.35
$\vdots$	$\vdots$	$\vdots$	$\vdots$	$\vdots$	$\vdots$	$\vdots$	$\vdots$
mean	72.3	1.94	0.31	$\cdots$	0.05	0.01	0.36

Fig. 6.1 shows the unweighted and weighted LRAs of this data set. The dimensionality of the data set is one less than the number of columns, equal to 10, so each

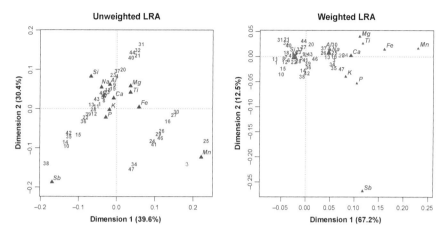

Fig. 6.1 Unweighted and weighted LRAs of the `RomanCups` data set. The symbol sizes are related to the weights given to the oxides, equal on the left, proportional to mean percentage on the right. Note the much higher variance in the unweighted analysis, caused mainly by the large differences in the logratios of *Mn* (see Table 6.2).

of these displays reduces the dimensionality from 10 to 2. Even though the displays look so different, they are both two-dimensional views of essentially the same sets of points, but with different importances placed on the points in the process of dimension reduction. The unweighted LRA, with equal weights on the parts, shows three bands of sample points, which correspond exactly to the percentages of 0.01, 0.02 and 0.03 for *Mn*. This element is the most important contributor to the unweighted total variance, owing to the huge ratios it engenders — consider that 0.03 %, for example, is 3 times 0.01 %, whereas for the element *Si*, the high values of the order of 70 % have much smaller ratio differences between them. Another way of describing this problem is that the data for *Mn* are reported with a high relative error, partially due to their being reported to only two decimal points.

In the absence of having information about the relative error of measurement of the oxides, a reasonable default weighting system is the one proposed here, assigning weights to the oxides proportional to their mean values. This will give *Mn* a very small weight and *Si* a large weight. Since each oxide contributes additively to the total variance, as shown in (5.2), their contributions can be compared in the two alternative analyses — see Table 6.2. Notice that, even though *Si* receives a very

Table 6.2 Contributions to total variance.

	Si	*Al*	*Fe*	*Mg*	*Ca*	*Na*	*K*	*Ti*	*P*	*Mn*	*Sb*
				Unweighted logratio analysis							
Part weights	0.091	0.091	0.091	0.091	0.091	0.091	0.091	0.091	0.091	0.091	0.091
Variance contribns (%)	7.7	2.5	5.9	6.4	1.9	3.6	4.1	4.9	5.4	29.2	28.3
				Weighted logratio analysis							
Part weights	0.724	0.019	0.003	0.005	0.057	0.183	0.005	0.0007	0.0005	0.0001	0.004
Variance contribns (%)	16.9	4.6	5.5	6.0	24.3	22.6	4.5	0.8	0.6	0.6	13.5

high weight in terms of its average value in the data set, it is only third highest in its contribution to the total weighted variance. Notice too that *Sb*, with very low values and thus a low weighting, still has a fairly high variance contribution, as it did in the unweighted analysis, whereas *Mn*'s contribution to weighted variance is now very small.

6.3 Different biplot scaling options

The plots in Fig. 6.1 are called *symmetric maps*, terminology borrowed from correspondence analysis, which has a strong relationship to LRA (see Chap. 8). These show logratio distances between the samples as well as logratio distances between the compositional parts — hence, the rows and columns are both spatially mapped in the same way. There are several alternative ways of showing the joint display in the form of true so-called *biplots*. In a biplot, one set of points, usually the samples, is displayed spatially as in the symmetric map, whereas the variables are scaled slightly differently to define different directions, called *biplot axes*. The projections of the samples onto the biplot axes are approximation visualizations of the centred data values. In an LRA plot, however, it is not the reconstruction of the CLRs that is of interest, but rather the visualization of the logratios themselves. Let us return to the smaller `TimeBudget` example and consider this issue in the symmetric map compared to the *asymmetric biplot*.

Fig. 6.2 shows the two options side by side. Notice that the asymmetric biplot has two scales this time, one for the samples and another for the parts, to accommodate the fact that the samples have a much smaller configuration than the parts. In practical terms, the difference between them is that in the symmetric map, both samples and parts have the same explained variances along the two dimensions of the solution, and can thus share the same scale on the axes. These common values are expressible as the same percentages of total variance shown on the two axes: of the total weighted variance of 0.03339, the amounts 0.02545 and 0.00517 are explained by the two axes, i.e. 76.2% and 15.5% respectively, which is true for both the row and column points in the symmetric map. Notice the greater dispersion of both sets

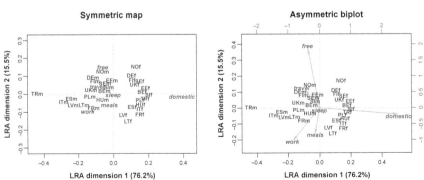

Fig. 6.2 Symmetric map and asymmetric biplot of the `TimeBudget` data set. Note the separate scale for the column points (in red) in the asymmetric biplot, since they have much higher spread than the row points (in blue). For abbreviations see Table 5.1.

of points in the symmetric map along the first axis compared to the second, which reflects these two quite different percentages. The symmetric map shows both rows and columns in *principal coordinates*.

In the asymmetric biplot the positions of the country/gender points are identical to those in the symmetric map, i.e. they are again shown in principal coordinates. The percentages on the axes refer to these row points only. The activity points, on the other hand, are now in *standard coordinates*, which have the property that their (weighted) variances are equal to 1 along each of the axes. Hence, it can be seen in the asymmetric biplot that the degree of dispersion of the activity points is similar along the two axes.

In both the symmetric map and the asymmetric biplot the positions of the compositional parts are not of interest; the links between them are. For example, the line connecting *free* and *domestic* represents the logratio log(*free/domestic*) (i.e. the direction from *domestic* to *free*) or the logratio log(*domestic/free*) (i.e. the opposite direction from *free* to *domestic*) — cf. the graph representation of logratios in Sect. 5.1. The distinguishing feature of the biplot is that each link itself defines a biplot axis for the corresponding logratio of the two activities. The perpendicular projections of the country groups onto this axis are approximations of the respective logratio values. This is not possible, strictly speaking, in the symmetric map — for example, in the symmetric map, the link between *free* and *domestic* still has the same orientation as in the biplot, but is less angled. Only if the percentages of variance on the two axes were close to one another would the link have practically the same angle in both plots, allowing the symmetric map to have the biplot interpretation.

Another scaling option for the parts (columns) is in terms of *contribution coordinates*. In this alternative scaling the property that the directions linking parts represent logratios is sacrificed in favour of showing which parts are more influential in determining the low-dimensional solution. This is typically used when there are lots of parts, so this option is illustrated with the `FishMorphology` data set, which has 26 parts in the form of relative measurements. Fig. 6.3 shows the asymmetric and contribution biplots. In this particular case there are no big differences

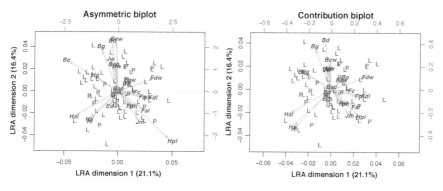

Fig. 6.3 Asymmetric biplot and contribution biplot of the `FishMorphology` data set. Fish are labelled L for littoral and P for pelagic. In both cases an additional scale for the variables (columns) is needed, since the fish "compositions" across the measurements have very low variance (the fish are all of the same species, *Arctic charr*).

(see the case study in Chap. 10 for more striking differences). But one can see, for example, that the measurements *Bc* and *Hpl*, upper left and lower right respectively, are not as important as they may appear to be in the asymmetric biplot on the left, having been "pulled in" by the contribution biplot.

6.4 Constrained compositional biplots

In Sect. 5.4 the full compositional data set was considered to constitute a set of responses, and its total logratio variance (see Sect. 5.3) was partially explained by a model in terms of some explanatory variables, using redundancy analysis (RDA). Using RDA for this objective is just a more efficient way of doing a bunch of multiple regression analyses, but RDA is also a dimension-reducing technique. Given several explanatory variables, RDA will isolate the space of the compositional data that is directly related to those variables, and then find the most important dimensions in this so-called *constrained* space. Hence, RDA performs an LRA (i.e. PCA of the CLRs) but in a part of the space with reduced total variance, which is just that part of variance accounted for by the explanatory variables.

The `FishMorphology` data set can illustrate the principle. A total of three explanatory variables were available: *Mass* (continuous variable), *Sex* (two categories) and *Habitat* (two categories). First, the two categorical variables are recoded into a single one with four categories, so that the sex–habitat interaction can be visualized. Mass, in this case log-transformed, and the sex–habitat categories define a four-dimensional subspace of of the data set: one dimension for log(mass) and three dimensions for the four-category variable, which is coded as four dummy variables, one of which is redundant. Of the total logratio variance of 0.001968 in the fish compositional measurements, only 0.000194 (9.9%) is explained by log(mass) and the sex–habitat variable with four categories. Both these variables are significant, as expected, according to the permutation test which is provided in the `vegan` package ($p = 0.002$ and $p = 0.01$ respectively).

The idea now is to restrict attention to that restricted variance of 0.000194, and see which ratios are associated with mass and the sex–habitat groups. Fig. 6.4 shows the RDA map containing three sets of points: the 26 relative measurements, the 75 fish as dots, and the explanatory variables displayed by the four group means and an arrow for *logmass*. The positions of the fish are determined by the four dummy variables and their masses — this explains why the four groups of fish lie in four bands, with the fish to the right in each band being the heavier ones. Notice that the ratio *Bcw/Fdw* coincides with the direction of *logmass*, which shows visually that this is an important ratio related to mass (see Tables 5.3 and 5.4). As mentioned before, conclusions about ratios should be made with care when the parts are displayed in contribution coordinates — it is preferable to do so when they are in principal or standard coordinates, as in the next plot (Fig. 6.5). As for the differences between the fish groups themselves, these are more or less in a direction orthogonal to the log(mass) vector, where pelagic fish are well separated from the littoral ones, with less separation between males and females within habitat groups. This again concords with the result found before that variable *Habitat* is more important than *Sex*.

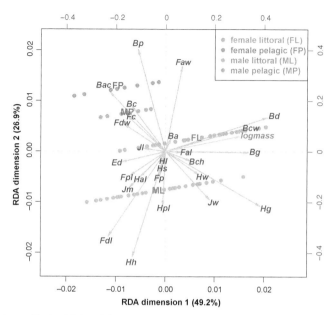

Fig. 6.4 RDA contribution biplot of the `FishMorphology` data set, constrained by the logarithm of *Mass* and the categorical variable coding *Sex* and *Habitat* interactively. The categories are at the averages of the fish groups.

Alternative ways of showing the fish are as supplementary points or as weighted averages and Fig. 6.5 illustrates the latter option. Each fish is shown at its weighted average positions of the measurements, which is the way correspondence analysis would situate them (see Chap. 8). A fish that has average values of all the relative measurements will lie exactly at the centre of the display, which is the centring condition on the measurement variables. As the shape of the fish changes, i.e. some relative measurements increase while others decrease, the fish's weighted average position relative to the measurement points migrates in one direction or another away from the centre. Thus the male littoral (ML) fish at the bottom of the display must have relatively higher values of *Hh*, for example, and lower values of *Bp* to be situated in that position. Because the fish are now distributed over the two-dimensional solution, 99% confidence ellipses for their mean points can be computed and shown — these show again that the littoral and pelagic groups are significantly separated, whereas sex differences appear non-significant, confirmed by permutation testing.

The constrained RDA space is four-dimensional and these figures show the best two-dimensional projection, accounting for 49.2 + 26.9 = 76.1% of that small part of the variance, which is in turn only 7.5% of the total variance of the data set. A permutation test for the four dimensions shows that the first two dimensions are significant ($p = 0.001$ and $p = 0.02$ respectively), but not dimensions 3 and 4. In fact, dimension 3 does oppose male and female fish, but the shape differences of these two groups have already been seen to be nonsignificant. This again supports the result that log(*mass*) and *Habitat* are the significant predictors of the fish shapes.

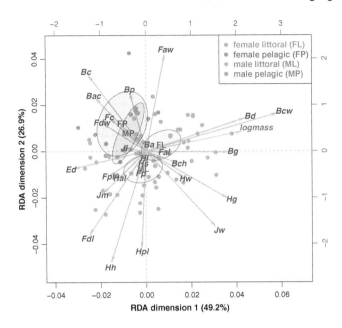

Fig. 6.5 RDA asymmetric biplot of the `FishMorphology` data set, constrained by log(*Mass*) and the *Sex–Habitat* combinations. Fish are situated at weighted averages of the standard coordinates of the 26 measurements; 99% confidence ellipses for the fish group means are added.

Summary: Dimension reduction using logratio analysis

- Logratio analysis (LRA) is a special case of weighted principal component analysis, where all ratios are analysed after log-transformation.

- LRA can equivalently be achieved by performing a weighted PCA on the CLR matrix, which is computationally more efficient. The default weights are the compositional part means; but, in principle, any set of user-defined weights can be used, and these can depend on the measurement error properties of the compositional parts.

- The reduced-dimensional LRA solutions have various scaling options. In the usual standard or principal coordinate scaling of the compositional parts, the links between the compositional parts in the LRA plot represent all the pairwise logratios. Alternatively, at the expense of losing this property, the parts can be displayed in contribution coordinates, to judge visually which parts are the most determinant in the LRA solution — this is useful when there are a lot of compositional parts.

- LRA can be constrained to depend linearly on a set of external explanatory variables, in which case the analysis is a redundancy analysis (RDA) of the CLR matrix. That part of the total variance of the original data set which is accounted for by the explanatory variables becomes the total variance of the RDA, to be decomposed along principal axes.

Chapter 7
Clustering of compositional data

Cluster analysis aims to form homogeneous groups of multivariate samples, called clusters, such that samples with similar sets of values are placed in the same cluster, and the clusters are as different as possible in terms of their constituent members. Two crucial decisions underlie any clustering algorithm: (1) how to define a measure of distance between the samples, which quantifies appropriately how "close" samples are to one another; and (2) how to define a corresponding measure of distance between the clusters. The special nature of compositional data and the distance functions discussed in previous chapters will prescribe the appropriate measures to use. Clustering can also be performed on the compositional parts themselves.

7.1 Logratio distances between rows and between columns

The logratio distance between samples (rows) has already been defined in the unweighted form (2.3). The more general weighted form, with weights c_j on the columns (compositional parts), is as follows, for the distance between two samples (rows) i and i':

$$d_{ii'} = \sqrt{\sum\sum_{j<j'} c_j c_{j'} \left[\log \frac{x_{ij}}{x_{ij'}} - \log \frac{x_{i'j}}{x_{i'j'}} \right]^2} = \sqrt{\sum\sum_{j<j'} c_j c_{j'} \left[\log \frac{x_{ij} x_{i'j'}}{x_{ij'} x_{i'j}} \right]^2} \quad (7.1)$$

For distances $d_{jj'}$ between parts (columns), simply interchange i and j and column weights c_j for row weights r_i in (7.1). Notice that the logarithm of the cross-product in the second formulation on the right of (7.1) remains unchanged — only the weights and the summation indices change to define the distance between columns, showing that the definitions of row and column distances are perfectly symmetric. The distance (7.1) can be equivalently and more conveniently computed using the CLRs $y_{ij} = \log(x_{ij}) - \sum_j c_j \log(x_{ij})$:

$$d_{ii'} = \sqrt{\sum_j c_j \left[y_{ij} - y_{i'j} \right]^2} \quad (7.2)$$

and similarly for the columns: first compute CLRs columnwise, then use the row weights and interchange row and column indices, thus changing the summation index from j to i.

7.2 Clustering based on logratio distances

Hierarchical cluster analysis can be performed on either the samples or the parts, using regular clustering algorithms, for example the function `hclust` in R. Using the
`TimeBudget` data and weighted logratio distances, Fig. 7.1 shows two complete
linkage clusterings, applied respectively to the demographic groups and to the activities. The dendrogram for the rows clusters all the female groups separately from
the male groups. The dashed line splits the male and female groups each into two
subgroups: for the females a separate quite homogeneous group is split off, mostly
of northern European countries (NOf, DEf, UKf, SEf, and FIf), and for the males a separate group of Latvia and southern European countries (LVm, ESm, ITm and TRm).
The dendrogram for the column clustering shows mainly a split-off of domestic activity from the others.

7.3 Weighted Ward clustering

The cluster analyses in Fig. 7.1 take row and column weights into account when
computing the logratio distances, but not in the actual clustering process. For example, even though each row is weighted equally, the fact that during the clustering
process some clusters have fewer members than others is not used in the definition
of between-cluster distance. Likewise, for the column clustering, some activities
have higher weights than others (in this case, higher averages) but only the distances between activities, not their respective weights, is used in the clustering. As

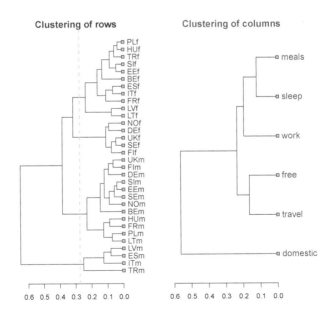

Fig. 7.1 Complete linkage clustering of rows (country–gender groups) and columns (activities) of
the `TimeBudget` data set. The dashed cutpoint in the left hand plot cuts the dendrogram into four
clusters. For row abbreviations see Table 5.1.

an alternative algorithm for clustering, the weighted form of *Ward clustering* can be used.

Ward clustering is based on minimizing the within-cluster variance and, equivalently, maximizing the between-cluster variance. Both of these variances use the weights of the objects in the definition of variance — taking into account these weights means that objects with lower weights are more easily absorbed into larger clusters. Since total logratio variance uses the weights of the rows or the columns, and since the within- and between-cluster variances will also take these weights into account, this criterion is a more natural one to use and has the benefit of decomposing the same total logratio variance into parts along the nodes of the dendrogram. The method needs a different algorithm, and is not just a simple application of the `hclust` algorithm in R using the option `method = "ward.D2"`, which gives regular Ward clustering. Fig. 7.2 shows the weighted Ward solution using an alternative R function `hierclust` provided as supplementary material. The dashed line for the row clustering cuts off the same clusters as in Fig. 7.1, and the column clustering shows the same contrast between domestic and the other activities. The difference now is that in both clusterings, the sum of the node heights is equal to the total logratio variance of the data set. For example, for the column dendrogram there are five nodes and the sum of the node heights is (starting from the lowest node where *sleep* and *travel* are joined, to the highest where *domestic* joins the others):

$$0.00103 + 0.00116 + 0.00365 + 0.00461 + 0.02294 = 0.03339 \qquad (7.3)$$

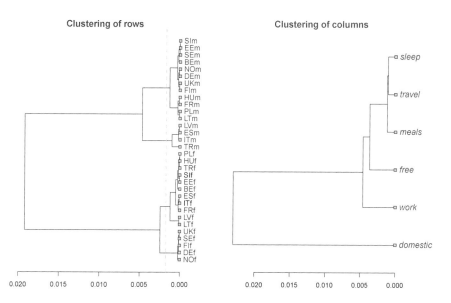

Fig. 7.2 Weighted Ward clustering of rows (country–gender groups) and columns (activities) of the `TimeBudget` data set. The dashed line in the left hand plot cuts the dendrogram into the same four clusters as in Fig. 7.1. For row abbreviations see Table 5.1.

The value 0.03339 in (7.3) is the total weighted logratio variance of this data set TimeBudget, as defined by (5.1) or (5.2). The same result is obtained by summing the 31 node heights for the row dendrogram. This property puts the clustering on a complementary footing with the logratio analysis (LRA). LRA decomposes the total logratio variance along principal axes, for either the row or column configuration (in principal coordinates), whereas Ward clustering decomposes the same total variance along the nodes of either of the row or column dendrograms.

The specific way the weighted Ward clustering proceeds is as follows, illustrated for the identification of the first two nodes of the column clustering. The first step is to join *sleep* and *travel*, with respective weights of 0.34125 and 0.05671, and a logratio distance of 0.14562 apart. They are joined together because they minimize the weighted Ward criterion:

$$\frac{c_{J_1} c_{J_2}}{c_{J_1} + c_{J_2}} \times d^2_{J_1, J_2} \tag{7.4}$$

where J_1 and J_2 are in general the two clusters being joined (in this initial step, just one column each), c_{J_1} and c_{J_2} are the respective weights, and d_{J_1, J_2} is the logratio distance between them. For this pair of activities, the value of (7.4) is:

$$\frac{0.34125 \times 0.05671}{0.34125 + 0.05671} \times 0.14562^2 = 0.00103$$

which is the first node height in the sum (7.3). Notice that the squared distances $d^2_{jj'}$ at the start of the clustering process are identical to the variances $\mathrm{var}_{jj'}$ defined in Sect. 5.3 (see page 38).

The way the two activities are now merged to proceed to the second step is not a simple aggregation of the *sleep* and *travel* values in the original table (see Sect. 7.4 where this possibility is considered). It is rather the weighted averaging of the CLRs of the two activities — these are the columnwise CLRs, not the rowwise ones used in (7.2): i.e., $z_{ij} = \log(x_{ij}) - \sum_i r_i \log(x_{ij})$, where the r_is are all equal to $1/I$, i.e. $1/32$ in this example. The new column *sleep&travel* has values $(c_j z_{ij} + c_{j'} z_{ij'})/(c_j + c_{j'})$, where $j = 3$ (*sleep*) and $j' = 4$ (*travel*) and $i = 1, 2, \ldots, 32$. The weight associated with this new column is the sum of the weights of the merged activities: $c_j + c_{j'} = 0.34125 + 0.05671 = 0.39796$. The distances between this new column and the others has to be recomputed and then the minimum of the same Ward criterion (7.4) is found. This turns out to be joining *meals* (weight 0.09747) with the new cluster *sleep&travel* (weight 0.39796), the distance between them being equal to 0.12165. The weighted Ward criterion is thus:

$$\frac{0.39796 \times 0.09747}{0.39796 + 0.09747} \times 0.12165^2 = 0.00116$$

which is the second node height in the sum (7.3). An equivalent way of computing and thinking about the merging of the columns is to make a weighted average of the log-transformed data: $\left(c_j \log(x_{ij}) + c_{j'} \log(x_{ij'}) \right) / (c_j + c_{j'})$ (which is the logarithm of the weighted geometric mean), then transform to the CLR of this new column.

The cluster analyses of the rows and the columns shown in Figs 7.1 and 7.2 do not give any information about why the clusters are formed. This type of interpretation is facilitated by the LRA of the data set, shown in Fig. 7.3. The distance of domestic activity on the right from all the others is clearly seen, with the female groups all on the right and the male groups on the left. The male groups that split off in the cluster analyses are at extreme left, spending the least time on domestic activity and the most on work activity. The mostly Nordic female groups that split off are at top right, having the most free time and the least work activity, especially Norwegian females (this can be verified in Table 5.1). Among the male groups, Norwegians similarly have the most free time.

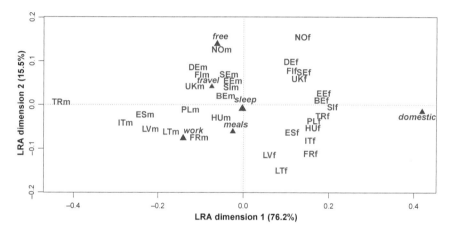

Fig. 7.3 Logratio analysis of `TimeBudget` data set, showing the symmetric map (repeated from Fig. 6.2 in larger format).

7.4 Isometric logratio versus amalgamation balances

As described in Sect. 3.4, a dendrogram can be used to define ILR balances, and these form a new set of $J - 1$ variables for a J-part composition. The theoretical advantage of these ILR balances is that they reproduce perfectly the logratio distances, while their disadvantage is the lack of an easy interpretation for the practitioner. So the question arises what the outcome would be to use simple aggregations instead of geometric means.

To answer this question in a specific example, consider the dendrogram of the six parts (activities) of the `TimeBudget` example in Fig. 7.2. This dendrogram generates five ILR balances of the form (3.5), but using weights according to our present approach. Hence, instead of $\sqrt{J_1 J_2 / (J_1 + J_2)}$ as the normalizing scaling factor in (3.5), $\sqrt{c_{J_1} c_{J_2} / (c_{J_1} + c_{J_2})}$ is used, where c_{J_1} and c_{J_2} are the accumulated weights of the corresponding sets of parts. Furthermore, the weights are also used in computing the geometric means. (Notice that the balances defined in (3.6) are for the complete linkage clustering of Fig. 3.1, whereas here it is for the weighted Ward clustering,

but this does not affect the argument.) If distances are now computed between the
32 rows of the data matrix using the ILR balances and compared with the logratio distances, the agreement is exact, as predicted by the theory: see the left hand
scatterplot in Fig. 7.4, showing the perfect agreement between the $32 \times 31/2 = 496$
interpoint distances computed in the two different, but equivalent, ways.

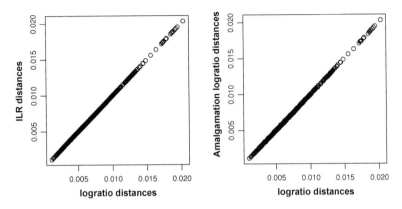

Fig. 7.4 Comparison of logratio distances and distances based on ILR balances (on the left) and
amalgamation balances (on the right), for the TimeBudget data.

On the other hand, instead of computing the geometric means of the ILR balances, suppose that the simple aggregations of the corresponding parts are formed
and expressed as logratios. For example, for the split of *domestic* versus the other
five activities, the ILR balance would compute the ratio of *domestic* and the geometric mean of the others. But now we rather compute the ratio of *domestic* to the
arithmetic sum (i.e. amalgamation) of the other five (which in this particular case
is reminiscent of an odds ratio $p/(1 - p)$), and then log-transformed). And once
again, the same scaling factor will be used as in the case of the ILR balance, but the
weighted form as described above. Having computed the five scaled *amalgamation
logratios* in this way, the inter-row distances are again computed and compared to
the original logratios, shown in the right hand plot of Fig. 7.4. The agreement is almost perfect, with a correlation of 0.99995 and even an agreement in scale. Whereas
the total logratio variance of the data set is equal to 0.033394, identical for the ILR
balances, the variance using the amalgamation logratios turns out to be 0.033358.

The amalgamation logratio deserves the title of a "balance" more than the ILR
one, since it is balancing real aggregations of parts rather than geometric means —
hence it will sometimes be termed an *amalgamation balance*. The amalgamation
balances almost perfectly reproduce the ILR balances, but with translation of scale,
as shown in Fig.7.5 for the first two balances: first, *domestic* versus the five others,
and second, *work* versus the four others (excluding *domestic*). Notice that the range
widths of each pair of balances are practically the same. And, as far as the translation

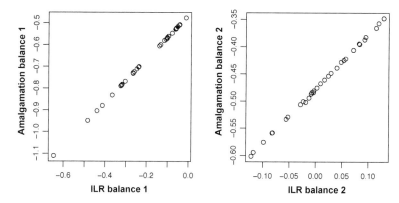

Fig. 7.5 Comparison of first two ILR and amalgamation balances, for the `TimeBudget` data.

of scale is concerned, remember that the amalgamation balance contrasts the sums of the parts, not their averages.

Admittedly, this is a small and uncomplicated data set, where in each balance one of the parts of the ratio is a single compositional part, so the question arises whether the same would occur when the numerators and denominators consist of many parts. So a new data set is introduced here, called `FattyAcids`, consisting of 27 fatty acid percentages in 52 samples of marine crustaceans of the amphipod family. This data set contains more than 100 zero values, which constitute a problem for the logratio transformation. These zeros were provisionally replaced by small positive values, a subject to be discussed in Chap. 8.

An important grouping of fatty acids is that of saturated, monounsaturated and polyunsaturated fatty acids, abbreviated as SFA, MUFA and PUFA respectively. These three major categories cover the complete set of fatty acids, and are routinely computed by amalgamating their respective constituents. In fact, ratios between these amalgamations are often computed, for example the key ratio PUFA/SFA.

Based on the two nested splits, (i) MUFA versus (PUFA and SFA), and (ii) PUFA versus SFA (to accommodate the ratio PUFA/SFA mentioned above), ILR and aggregation balances computed on the 52 samples are compared in Fig. 7.6. The agreement is far from the almost exact one observed in Fig. 7.5, but nevertheless the relationships are largely monotonically increasing.

As a final commentary, the three major groups, SFA, MUFA and PUFA, form a three-part composition, which defines logratio distances between the samples. The two amalgamation balances computed on the full data set perform better at reproducing these distances than the corresponding ILR balances, as can be seen in the two plots of Fig. 7.7 — the nonparametric Spearman rank correlation coefficient for the scatterplot in the left hand plot is 0.996 whereas in the right hand plot it is 0.979. This is expected since these three groups are defined as sums of fatty acid parts (i.e. amalgamations) by biochemists, never by computing geometric means. The ILR balances are additionally affected by the relative values of the parts constituting the geometric means.

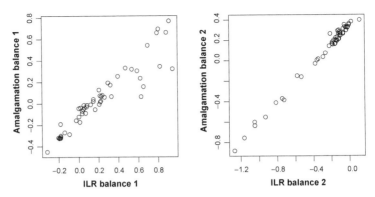

Fig. 7.6 Comparison of the ILR and amalgamation balances, based on three subsets, in the FattyAcids data set: monounsaturated (MUFA), polyunsaturated (PUFA) and saturated (SFA) fatty acids. The first balance is MUFA versus the rest, and the second is PUFA versus SFA.

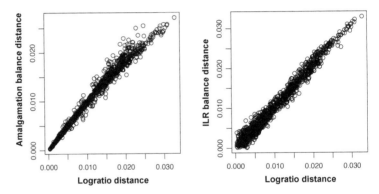

Fig. 7.7 Comparison of inter-sample distances based on amalgamation and ILR balances, with the logratio distances of the three-part composition formed by SFA, MUFA and PUFA fatty acids.

Summary: Clustering of compositional data

- The logratio distance, usually weighted, is the natural distance measure for clustering rows or columns of a compositional data set.
- Any preferred clustering criterion can be used, but Ward clustering (weighted or unweighted), decomposes total logratio variance (weighted or unweighted) along the nodes of the cluster dendrogram.
- The nodes of a dendrogram can define isometric logratio (ILR) balances. However, amalgamation balances, formed as logratios of amalgamated parts that contrast groups of parts, show acceptable performance in practice and can serve as more interpretable alternatives to ILR balances.

Chapter 8
Problem of zeros, with some solutions

Zero values are often present in a compositional data set, and these are problematic for computing ratios. Zeros are often replaced by small positive values, where the strategy for choosing such values depends on the nature of the compositional data. An alternative is to use a different transformation of the data, which accommodates zeros and is "close" in a certain sense to the logratio one, thus approximating the logratio structure. Closeness can be measured, for example, by how far away a transformation is from being subcompositionally coherent, i.e. its level of subcompositional incoherence. The chi-square distance in correspondence analysis is a possible alternative, often with low subcompositional incoherence, since logratio analysis and correspondence analysis have strong theoretical and practical connections.

8.1 Zero replacement

Apart from the `FattyAcids` data set used in Chap. 7, there have been no zero values in the compositional tables treated so far. But zeros frequently occur in compositional data, for many reasons. The proportions could have been based on counts, where several counts were zeros. Or the data might be rounded to two decimals, say, and several values appear in the reported data table as 0.00, when they were really a small positive amount. Whatever the reason for the occurrence of zeros, it is clearly not possible to apply the logratio techniques described in this book, since division by 0 is not possible, nor can log(0) be computed.

A common-sense strategy, but rather ad hoc, is simply to replace the zeros with small positive values. One option is to identify, for each compositional part, the smallest positive value, and then replace the zeros for the respective part by a fraction (e.g. a half) of that small value. If the data had been rounded to two decimals, for example, and no other information is available except it is known that zero values are impossible, the zeros could all be replaced by 0.005. In biochemical and geochemical studies, the so-called *detection limit* for each part might be known and these limits could be different for each compositional part. Then a fraction of that limit could replace a zero for the respective part, and again half the detection limit seems a reasonable choice.

Whatever zero replacement strategy is adopted, the addition of these new values to the data set affects the row margins of the table, which are no longer constant, so the rows need to be reclosed as proportions or percentages. It might be of concern, especially if many zeros have been replaced, that these changes in the data might substantially affect the results. Then a *sensitivity analysis* can be conducted, to try several choices of the zero replacement values and see the effect on the logratio structure.

8.2 Sensitivity to zero replacement

In the previous chapter, the 52×27 `FattyAcids` data matrix had 104 zeros, which represent 7.4% of the data values. Sixteen of the 27 columns (fatty acids) included some zeros, varying from one zero in a column to 19 zeros (in column 2, for fatty acid *15:0*). To be able to analyse these data in Chap. 7, the zeros were simply replaced by a half of the smallest positive value in each column. To get an idea how this decision affects the result, the possibility that zeros were replaced by values as low as a tenth of the smallest positive value are also considered. Thus 100 alternative data sets are generated, with the zeros replaced by a random value between a tenth and a half of the smallest positive value in the respective column. Fig. 8.1 shows the original LRA biplot, using the contribution coordinates to scale the fatty acids, and showing only those fatty acids that contribute more than average to the principal axes. Under each fatty acid label the number of zeros is given in parentheses, and similarly for the samples, which are displayed as light blue dots. The fatty acid *20:4(n-3)* at bottom left has nine missing values and there are nine samples on the right (with ID numbers above their dot symbols) that show the replicated sample

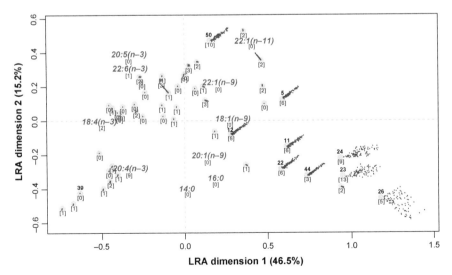

Fig. 8.1 Logratio contribution biplot of `FattyAcids` data set, displaying sample points as light blue dot symbols and additional points for each sample as tiny blue dots after randomizing the zero replacement values 100 times. The number of zeros for each displayed fatty acid is given in parentheses, and the ID numbers of nine samples on the right are shown above their dot symbols (these are referred to in the text).

points diverging away from *20:4(n-3)* towards top right. As the replacement value for this fatty acid is decreased from a half to a tenth of its minimum positive value, the repeated points drift further to top right. Samples labelled 24, 23 and 26 at bottom right show, in addition to this dispersion away from *20:4(n-3)*, two different patterns: samples 23 and 24 show dispersion to the top right as well as to the right, both of them having the zeros of *18:4(n-3)* on the left side of the horizontal axis, and sample 26 has the only zero of *22:6(n-3)*, which is top left, resulting in its replicate points having a parallelogram shape, away from both *20:4(n-3)* and *22:6(n-3)*. Using the contribution biplot is useful in this case, because it shows only the most determinant fatty acids in obtaining the solution. Notice that fatty acid *15:0*, which had 19 zeros, does not appear in Fig. 8.1, not being one of the high contributors to this solution, hence varying its zero replacement value would hardly affect the positions of the samples.

All in all, the results in Fig. 8.1 show a strong degree of stability in terms of the distance structure between the samples, under a wide range of small zero replacement values. A similar style of sensitivity analysis can be conducted depending on the nature and origin of the data. For example, when the compositions are based on count data and zero counts are present, it is customary to add 1 to all the data values, and then form the compositions before log-transforming. The sensitivity of the results can be studied by adding a series of smaller values, down to 0.1, say.

8.3 Subcompositional incoherence

Subcompositional coherence (see Sect. 1.4) is one of the most important properties governing compositional data analysis. But let us suppose that there is an alternative approach that is not coherent, i.e. incoherent, but has other favourable properties, for example it allows zero values. If this lack of coherence can be measured and the alternative method, applied to a specific data set, turns out to be quite close to being coherent, then the method can still be justifiably applied. What is needed then is a measure of *subcompositional incoherence*.

The principle of subcompositional coherence is aimed at the compositional parts of a data matrix, i.e. the columns. This principle states that the structure of a subset of parts is not affected by forming a subcomposition of those parts. Since the structure of the parts can be measured by their between-part distances, these distances can form the basis of measuring subcompositional incoherence.

In order to study alternative distance measures, suppose there is a measure between pairs of the J parts that leads to a $J \times J$ matrix Δ of distances $\delta_{jj'}$. There are a multitude of possible subcompositions of J parts, but the "worst case scenario" is that of a two-part subcomposition, i.e. choosing two parts, and then closing them sample-wise. This can be done for each of the $J(J-1)/2$ pairs of parts, and the same alternative distance function applied to compute the distance between the two members of the pair. If all these pairwise distances are assembled in another square symmetric matrix, denoted by $\tilde{\Delta}$, then the two matrices Δ and $\tilde{\Delta}$ can be compared to see how similar they are. If the distance function is the logratio distance, we know that — thanks to subcompositional coherence — the two distance matrices will be identical. But for any other distance measure, there will be a difference between the two matrices. In multidimensional scaling (MDS) there is a standard way to measure the difference between two distance matrices, called the *stress*, so this can be conveniently used here. The general weighted form is preferred, which places more importance on distances between "abundant" parts, but the unweighted form can always be obtained by setting the weights c_j equal to $1/J$:

$$\text{stress} = \sqrt{\frac{\sum_{j<j'} c_j c_{j'} (\delta_{jj'} - \tilde{\delta}_{jj'})^2}{\sum_{j<j'} c_j c_{j'} \delta_{jj'}^2}} \qquad (8.1)$$

This stress measure was computed for several distance measures in the compositional data analysis literature, for the `FattyAcids` data set, listed here in ascending order of subcompositional incoherence:

Distance measure	Incoherence
Logratio	0
Chi-square	0.194
Euclidean	0.396
Hellinger	0.628
Arc sine	0.739

The stress for the logratio distance is 0, as expected, since the distance is subcompositionally coherent. The chi-square distance has relatively low incoherence, while all the other measures are further away from being coherent. It seems that the chi-square distance could be a good alternative to the logratio distance, at least for this data set.

8.4 Correspondence analysis alternative

The chi-square distance underlies correspondence analysis (CA), which is a competitor to logratio analysis in that it also analyses relative values in data matrices of nonnegative values. "Nonnegative" is the key word, because CA can analyse tables with many zeros and is the multivariate dimension-reduction method of choice for ecologists, linguists and archaeologists, who regularly have very sparse data tables, i.e. tables with many zeros. CA also relativizes the data values, not by taking ratios but by expressing rows and columns relative to their respective sums, called row and column *profiles* in the terminology of CA. Profiles are none other than compositions, row- and column-wise.

The two-dimensional CA biplot comparable to that of Fig. 8.1 is shown in Fig. 8.2, with the same samples labelled as before, for comparison. First, the major contributing fatty acids overlap with those in Fig. 8.1 but differ in a few cases. The configuration of samples resembles the one before, but the samples at bottom right do not extend out as much as before. A way of measuring the similarity between the two configurations is to use the correlation from a *Procrustes analysis*, available in the R package `vegan`. This correlation between the sample configurations

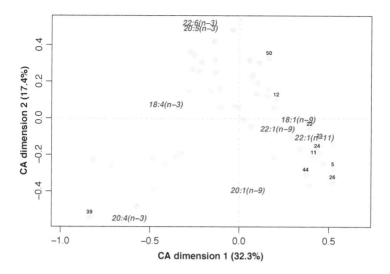

Fig. 8.2 CA contribution biplot of `FattyAcids` data set, including all the original zeros. The fatty acids contributing more than average to the principal axes are shown. Sample labels are shown for the same 9 samples labelled on the right of Fig. 8.1.

in Figs 8.1 and 8.2 is equal to 0.869. By computing inter-sample distances in both two-dimensional solutions and making a scatterplot of them, this correlation can be better appreciated — see Fig. 8.3. The positive, but not very strong, relationship with the LRA distances can be improved using a power transformation.

CA and the chi-square distance have a close theoretical relationship with LRA and the logratio distance, via the Box-Cox power transformation:

$$f(x) = \frac{1}{\alpha}(x^{\alpha} - 1) \tag{8.2}$$

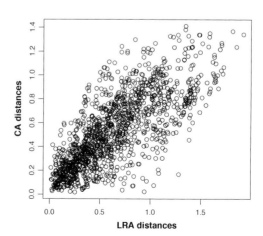

Fig. 8.3 Logratio distances between samples from the two-dimensional solution in Fig. 8.1 (the light blue dots) compared to those from the CA solution in Fig. 8.2 .

The power α is usually larger than 0 and less than or equal to 1, and as α tends to 0 the transformation tends to the log-transformation. There are two ways this transformation can be applied: one links CA directly to unweighted LRA and the other to weighted LRA. In both cases, the term -1 in (8.2) is not necessary, since it will be eliminated by the double-centring inherent in CA. The first way is simply to apply (8.2) without the -1 term, i.e. $f(x) = (1/\alpha)x^{\alpha}$ to the original compositional data set $[x_{ij}]$, for a series of CAs starting from the transformed data with α equal to 1 and descending to the transformation with a very small positive value. At the start of this series the analysis would be regular CA, with chi-square distances, and at the end the analysis would tend to an unweighted LRA, with unweighted logratio distances. The second way is to perform the power transformation on the ratios x_{ij}/c_j, where c_j are the column weights — in the terminology of CA, these are called *contingency ratios*, the compositional values relative to their expected values (i.e. means). In this case the series of analyses starts with regular CA and tends in the limit to weighted LRA, with weighted logratio distances, as α tends to 0. This is a very convenient result, since it means that LRA can be approximated as closely as one likes by a CA of a suitably power-transformed compositional data matrix.

This idea can be illustrated in the latter weighted case by using, for example, $\alpha = 0.5$ (i.e. square root transformation), and seeing the outcome in the CA. Fig. 8.4 shows the result, and the map is already starting to resemble Fig. 8.1 more closely — notice how sample 26 is moving away from the others and the improved positional similarity of the higher contributing fatty acids.

If the data were all positive, with no zeros, reducing α to near 0 would give almost exactly the same as the LRA solution, as the transformation approaches the

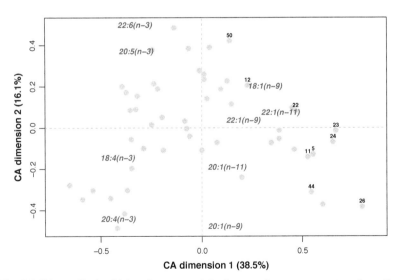

Fig. 8.4 CA contribution biplot of `FattyAcids` data set, after square root transformation of the contingency ratios. The same subset of sample labels has been added as in previous figures, for purposes of comparison.

log-transformation. However, the presence of zeros will mean that at some point the reduction of α will start to degrade the solution. Procrustes correlations and measures of stress can help to find out what α should be used to optimize the CA solution to the LRA one. Here the stress formula (8.1) is used for a different purpose, to measure the lack of agreement between inter-sample distances in a power-transformed CA compared to LRA, as the power changes.

Hence, a series of values of α is tried, $\alpha = 1, 0.99, 0.98, \ldots, 0.01$ and in each transformed data set both the Procrustes correlation between the power-transformed CA and the LRA as well as the stress (8.1) are computed, for the full space solution as well as the two-dimensional solution. The Procrustes correlation will increase and stress will decrease for decreasing α, and would end up at 1 and 0 respectively if there were no zeros present, but with zeros in the present FattyAcids data set, both measures reach a turning point — Figs. 8.5 and 8.6 show the results.

Notice first in Fig. 8.5 how the curves in red, corresponding to CA on the data with no zeros (i.e. zeros replaced by small positive values), continue increasing towards a correlation of 1 as α decreases from right to left. Similarly, in Fig. 8.6 these red curves continue decreasing towards a stress of 0 as α tends to 0. Both

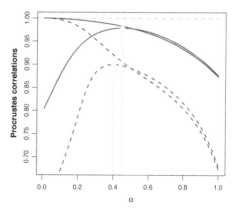

Fig. 8.5 Procrustes plots: solid curves — for full space solutions, the correlation with the LRA solution of CA of the data set with zeros (blue curve) and the data set with zeros replaced (blue curve); dashed curves are for the two-dimensional solutions. Maxima are indicated on the blue curves.

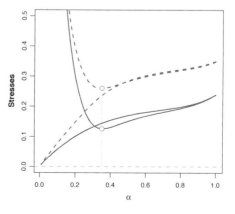

Fig. 8.6 Stress plots: solid curves — for full space solutions, stress between the LRA solution and the CA of the data set with zeros (blue curves) and the data set with zeros replaced (red curves); dashed curves are for the two-dimensional solutions. Minima are indicated on the blue curves.

confirm the theoretical result that with this power transformation, CA on a strictly positive data set tends to LRA, equivalently the chi-square distances tend to the logratio distances, as the power tends to 0.

Reading the blue curves from right to left as α decreases, corresponding to CA of the data with zeros, Fig. 8.5 shows an increasing correlation reaching an optimum highest correlation at $\alpha = 0.45$ for the full space configuration, and $\alpha = 0.40$ for the two-dimensional solution. Similarly in Fig. 8.6, the stress decreases to a minimum at $\alpha = 0.35$, for both the full and reduced configurations. This adds some credence to the choice of the square root transformation that was used to obtain Fig. 8.4 — the value 0.5 for α shows values for the Procrustes correlation and stress quite close to their optima. The value of subcompositional incoherence for $\alpha = 0.5$ is 0.0936, less than the value of 0.194 for CA on the original untransformed data ($\alpha = 1$).

Thus, when there are zeros present in the data, zero replacement can be avoided by using CA, with the option of reducing the subcompositional incoherence by introducing an appropriate power transformation, either of the original compositional data values (for an unweighted analysis) or of the contingency ratios (for a weighted analysis).

Summary: Problem of zeros, with some solutions

- The use of the logratio transformation implies that the data set should consist of strictly positive values. But many data sets do contain zeros.
- One strategy is to replace the zeros with some small positive number — this approach will be governed by how the data were collected, whether the zeros are real zeros, from data rounding, or a result of being lower than the detection limit.
- Whatever the replacement approach is, it is advisable to conduct a sensitivity analysis, varying the replacement values within reasonable limits to see the effect on the data analysis, results and interpretation.
- CA is a competitor to LRA, and can analyse data with zeros, but it is not subcompositionally coherent. Its lack of coherence, i.e. subcompositional incoherence, is seen in general to be lower than other multivariate methods that are routinely applied to compositional data.
- For strictly positive data, CA tends to LRA, and thus tends to subcompositional coherence, when the data, or contingency ratios, are power transformed.
- This means that CA is a good alternative to LRA when there are many zeros in the data, and these zeros do not need replacement.
- Because of CA's theoretical connection to LRA, the application of CA to compositional data can be further enhanced by introducing a power transformation to further reduce its subcompositional incoherence.

Chapter 9
Simplifying the task: variable selection

Faced with a complex compositional data set with many components, a major simplification can be achieved by trying to identify which logratios are the most important drivers underlying the data structure. This subject has been partially addressed in Chap. 5, where logratios were identified that were the most significant predictors of a continuous or categorical response variable. In the present chapter, however, the whole compositional data set is considered as a response set, and logratios constructed from the same data set itself are sought that explain a maximum part of the total logratio variance. The result is that the original data set can effectively be replaced by a small set of logratios, which can then be treated as regular interval-scale statistical variables. Since this selected set of logratios involves a subset of the compositional parts, this approach can also be considered a way of choosing the most important subcomposition underlying the data.

9.1 Explaining total logratio variance

The total variance of a J-part compositional data set can be quantified in several equivalent ways; for example, as the (weighted) average of all $J(J-1)/2$ pairwise logratio variances, or the (weighted) average of all J centred logratio variances (CLRs). It is known that the dimensionality (i.e. rank) of such a data set is equal to $J-1$, so amongst the pairwise logratios only $J-1$ independent logratios are required, and all the other $\frac{1}{2}J(J-1)/-(J-1) = \frac{1}{2}(J-1)(J-2)$ logratios will depend linearly on them.

Returning to the `RomanCups` archaeometric data set introduced in Chap. 6 (the 11-part oxide compositions of 47 Roman glass cups), there are 55 logratios, but only 10 of them are needed to generate the whole data set. Such a subset of 10 logratios is not just any subset, the chosen ratios must themselves have no dependencies among them; that is, they should form a graph that is both acyclic (i.e. no cycles) and connected (i.e. all vertices in the graph are joined by edges — see Sect. 5.1). Fig. 9.1 shows first the graph of all logratios, then the graph of a set of 10 additive logratios (ALRs), which is just one of the many possible acyclic connected graphs, and finally another particular acyclic connected graph involving a selection of logratios that will soon be explained. The second graph is shown with arrows connecting the vertices, i.e. a directed graph, whereas the other two are undirected, but this is not important

Fig. 9.1 Left: graph of all pairwise logratios between the 11 parts. Middle: graph of 10 additive logratios, with *Si* as the denominator. Right: a particular acyclic connected graph of 10 logratios. The first graph is shown as undirected, the second as directed and the third undirected but where the edge widths are related to the stepwise process of logratio selection (Sect. 9.2, Table 9.2).

to our discussion here, since reversing the direction just inverts the ratio (i.e. changes the sign of the logratio).

From graph theory, there are J^{J-2} different acyclic connected graphs, which in this case is $11^9 = 2\,357\,947\,691$, over 2 billion. The set of 10 logratios defined by any one of these would explain all the total logratio variance, but one of these can be efficiently chosen with some other favourable properties. For example, the middle graph in Fig. 9.1 shows one possible set of additive logratios (ALRs), of ten of the oxides referred to the oxide of silicon (*Si*), i.e. silicon dioxide, or silica. There are 10 other sets of ALRs possible, depending on the reference oxide, but the ALRs relative to *Si* contain (or contribute) the highest amount of the total variance out of the 11 possible sets of ALRs (see Sect. 9.2 and Table 9.1). A distinction needs to be made between "variance explained" (in the regression sense) and "variance contained" (or contributed, in the sense of parts of variance). Each of the 11^9 acyclic, connected sets composing 10 logratios will explain 100 % of the variance. But each set will contribute only 10 terms of the variance of the 55 logratios, the other 45 components coming from logratios that are linearly dependent on these 10. These parts of contributed variance, or variance contained, will be different in each case. In what follows, however, interest will be more focused on explained variance.

9.2 Stepwise selection of logratios

In Sect. 5.4 redundancy analysis (RDA) was used to model the complete compositional data set on several additional predictor variables. The response set is conveniently defined as the set of CLRs, equivalent to using the much larger set of logratios. The same approach is now followed to model the compositional data set, but using the logratios themselves as explanatory variables. This is not as strange as it might seem — a particular logratio certainly explains (and contains) its own variance, which is just one term in the total logratio variance, but it will inevitably be correlated with many other logratios and so explain parts of their variances too.

A simple strategy is to limit the search to the different sets of ALRs, since there are only 11 possible sets. Each set explains 100% of the logratio variance, but some sets contain far more variance than others. Table 9.1 shows the ALR sets in descending order of contained variance. The weight of each oxide is also given as well as the Procrustes correlation between the 10-dimensional configuration of the samples (Roman cups) in the LRA (i.e. the full space of all logratios) and the 10-dimensional configuration based on the corresponding set of ALRs. The contained variance (i.e. part of variance contributed to total variance) more or less follows the order of the weight given to the reference oxide, which is to be expected, since the total variance is weighted. The Procrustes correlation, on the other hand, can be quite high for ALRs with reference oxides such as *Mg* and *Al* that have low weights and low contributions to total variance. The clear "winner", however, is the ALR with *Si* as reference, which contains almost 90 % of the total variance and is practically identical to the logratio structure, with a Procrustes correlation of 0.998.

Table 9.1 ALR sets.

Reference oxide	R^2	% contained	Weight	Procrustes
Si	1	89.3%	0.7237	0.998
Na	1	40.9%	0.1825	0.844
Ca	1	30.0%	0.0567	0.699
Sb	1	13.9%	0.0036	0.769
Al	1	6.5%	0.0194	0.904
Mg	1	6.4%	0.0046	0.954
Fe	1	5.8%	0.0031	0.670
K	1	5.0%	0.0040	0.702
Ti	1	0.9%	0.0007	0.656
P	1	0.7%	0.0005	0.619
Mn	1	0.7%	0.0001	0.569

A more complex strategy is to conduct the search amongst single logratios, not sets of them. One logratio will explain the maximum amount of variance and that is the one that will be identified first — it turns out to involve the ratio of oxides *Si/Ca*, and this single logratio explains 61.5 % of the total variance.

The best ratio is thus fixed as an explanatory variable in the RDA and the next best one identified to explain a maximum of the residual variance — it is *Si/Sb*. So now there are two logratios fixed in the model, and a third one is sought with the same objective. This process will continue and have exactly 10 steps until 100% of the logratio variance is explained. The list of ratios entering, their additional and accumulated variance explained, as well as the variance contained, is given in Table 9.2. Notice that by the last step all variance is explained but only 49.7 % of the logratio variance is actually contained in the 10 ratios, the remaining variance being contributed by the other 45 logratios that depend linearly on the chosen 10.

These 10 ratios also define a sample configuration in a 10-dimensional logratio space, and the Procrustes correlation with the configuration defined by all the logratios is 0.936. This is a correlation between configurations in the full space of the logratio structure, but generally the low-dimensional projection onto the plane of the first two dimensions would be interpreted. The two planar projections are shown in comparable contribution biplots in Fig. 9.2. The sample points have a clear

Table 9.2 Oxide ratios entering stepwise as logratios in explaining maximum logratio variance: additional percentage of total variance (which equals 0.002339) explained by the ratio, accumulated percentage, additional percentage of total variance contained, and accumulated percentage. Their graph is shown in the right hand plot of Fig. 9.1.

Oxide ratio	% variance explained	% variance explained (accum.)	% variance contained	% variance contained (accum.)
Si/Ca	61.53 %	61.53 %	25.82 %	25.82 %
Si/Sb	12.59 %	74.12 %	10.41 %	36.23 %
Na/Sb	12.30 %	86.42 %	2.33 %	38.56 %
Si/Fe	7.17 %	93.58 %	4.68 %	43.24 %
K/Sb	3.04 %	96.62 %	0.06 %	43.30 %
Si/Mg	1.83 %	98.45 %	5.09 %	48.39 %
Al/Na	0.75 %	99.20 %	0.74 %	49.13 %
Ca/Ti	0.31 %	99.51 %	0.03 %	49.16 %
Si/Mn	0.28 %	99.78 %	0.51 %	49.67 %
P/Sb	0.22 %	100.00 %	0.01 %	49.68 %

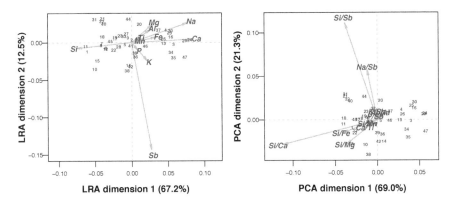

Fig. 9.2 Left: LRA contribution biplot of the `RomanCups` data set. Right: Contribution biplot of the weighted PCA of the 10 logratios listed in Table 9.2. In both plots variables near the centre make small contributions to the solution.

positional association when comparing the two plots and the Procrustes correlation is now equal to 0.963 in these planar projections, higher than the correlation in the full 10-dimensional spaces.

Notice that the percentages of explained variance on the axes of these two biplots are referred to different totals: in LRA the total is the total logratio variance, whereas in the PCA of the 10 logratios, the total variance is that of the 10 logratios only.

9.3 Parsimonious variable selection

The biplots of Fig. 9.2 are both contribution biplots, where the coordinates of the variables (compositional parts and logratios respectively) are related to their contributions to the two dimensions of the respective solutions. There are clearly some parts that stand out in the first biplot and, similarly, for the logratios in the second biplot. The question arises whether we need all the parts and all the logratios to describe the structure adequately. For example, in the biplot of the logratios there are five logratios that stand out clearly from the rest: *Si/Ca*, *Si/Sb*, *Na/Sb*, *Si/Fe* and *Si/Mg*. According to Table 9.2 these five logratios explain $61.5 + 12.6 + 12.3 + 7.2 + 1.8 = 95.4\%$ of the total variance and their parts of variance account for $25.8 + 10.4 + 2.3 + 4.7 + 5.1 = 48.3\%$ of the total. These logratios involve only six of the 11 oxides, and the question is whether the smaller set of five logratios, or the subcomposition formed by the six parts, is sufficient to describe the structure of this data set. Fig. 9.3 shows the respective biplots using just these reduced sets of parts and ratios respectively. Comparing these with the biplots in Fig. 9.2, there is no noticeable difference at all, so the reduced sets of parts or logratios serve the same purpose as the original sets. Compared to the LRA biplot in Fig. 9.2, the Procrustes correlations of the two biplots in Fig. 9.3 are 0.997 and 0.964 respectively.

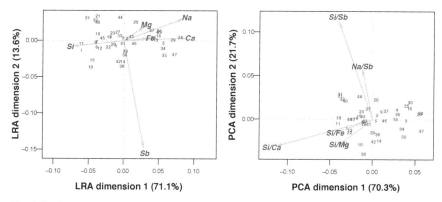

Fig. 9.3 Biplots of the `RomanCups` data set comparable with Fig. 9.2, but with parsimonious selection of parts and ratios. Left: LRA contribution biplot of the subcomposition of six parts. Right: PCA contribution biplot of five logratios. Apart from changes in the percentages on the axes, there are almost no discernible differences in the configurations of the sample points.

The sample configurations remain stable, but there are differences in the percentages on the axes, because in both cases the measures of total variance have changed. In the case of the LRA of the subcomposition, the total variance is now reduced to that of the subcomposition, while in the case of the PCA of the subset of logratios the total variance is reduced to that of the subset. Since some of the variance that can be regarded as random variation has been removed in these reduced sets of variables, the overall variances explained have increased, especially along the first principal axes. Notice too that in the case of the LRA of the subcomposition, the total variance includes the variance of all 15 logratios between the six oxides, whereas in the PCA of the logratios, the variances of only five of those 15 are included.

This parsimonious description of the structure of the compositional data set can be of substantial practical benefit to the practitioner. The small subset of logratios is especially useful, since they can be reported as valid univariate statistics, with their measures of centrality and dispersion like regular statistical variables. In fact, the ratios themselves, which tend to be skewed, can be summarized by their medians and reference ranges — these ranges are estimates of the 2.5 % and 97.5 % percentiles of their respective distributions (Table 9.3), using the `quantile` function in R. These summary statistics can be compared between different studies, archaeometric studies in this case, since they are subcompositionally coherent.

Table 9.3 Univariate statistics — medians and reference ranges — for the five selected ratios.

Ratio	Median	Reference range
Si/Ca	13.3	10.1 – 15.0
Si/Sb	206.5	120.4 – 403.5
Na/Sb	53.3	32.1 – 93.6
Si/Fe	244.3	163.8 – 340.8
Si/Mg	157.8	117.3 – 230.0

9.4 Amalgamation logratios as variables for selection

Amalgamations, if meaningful in the context of the data, can also be included in logratios for possible selection in explaining logratio variance. The only problem is a technical one, because the number of possible amalgamations is huge, if one were to conduct an exhaustive search through all the possibilites. Restricting attention to just two-part amalgamations, there are 55 possibilities, and then, combining these with the 11 original parts (i.e. "one-part amalgamations"), this leads to a very large number of possible logratios, $66 \times 65/2 = 2145$. Not all of these, however, are to be considered; for example, the numerator and denominator of each ratio should not have parts in common. The number of ratios is still manageable for this relatively small data set, so this "data mining" search was conducted to see how much it improves the variable selection.

Table 9.4 shows the results of a similar stepwise process as before, but including ratios of two-part amalgamations, in explaining the total logratio variance of the 11-part composition — notice that the total variance of the original composition is still taken as the target to be explained. Hence, in this case it is not possible to compute variance contained, since the amalgamation logratio variances are not included in the calculation of total variance. The parts of variance explained are improved compared to Table 9.2 (they could not be worse, since the logratios of single parts are also included in the search).

9.5 Signal and noise in compositional data

Using the 10 logratios in Table 9.4, Fig. 9.4 shows the PCA contribution biplot. For the first time, the configuration of the sample points shows some difference compared to the LRA one as well as the others shown before (Figs. 9.2 and 9.3). In particular, the importance of the second dimension is much reduced, and the first dimension captures 90.8% of the variance. Only the first four logratios show high contributions to the solution, whereas the other six show very small contributions since they lie practically at the origin of the biplot.

This result suggests a reconsideration of what is truly signal and noise in these data. All data contain random variation and throughout the recent analyses there

Table 9.4 Amalgamation ratios that enter the stepwise process of explaining maximum logratio variance of the original composition: additional percentage of variance explained by each ratio and accumulated percentage.

Amalgamation ratio	% variance explained	% variance explained (accum.)
Si/(Ca+Na)	64.54 %	64.54 %
Na/(Sb+P)	13.01 %	77.55 %
(Mg+Al)/(Na+Sb)	10.73 %	88.29 %
Na/(Ca+P)	6.00 %	94.28 %
(K+P)/(Mg+Sb)	3.04 %	97.32 %
(Al+Fe)/(Mg+Ti)	1.27 %	98.59 %
(Al+P)/(Fe+Mn)	0.55 %	99.15 %
(Fe+K)/(Ti+P)	0.36 %	99.51 %
Mn/(Fe+Ti)	0.26 %	99.77 %
Ti/P	0.18 %	99.95 %

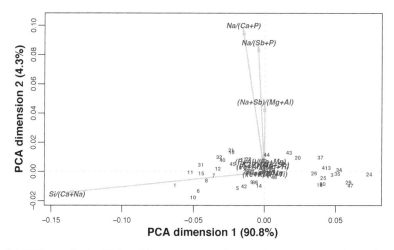

Fig. 9.4 PCA contribution biplot of the RomanCups data set where two-part amalgamations have been included in the search for ratios. The ratio *(Mg+Al)/(Na+Sb)* has been inverted here.

has been an implicit assumption that the total logratio variance is the quantity of interest to explain, as if it contained no noise. At least, it has been assumed that the first two dimensions of the LRA of the data are "real", in the sense of not being a result of random variation. There are several ways to check this assumption. One way is to generate permutation distributions for the variance components on the axes, by randomizing each set of oxide values several times (999 times in this case), and then seeing where the original observed parts of variance lie with respect to these respective distributions. It turns out that only the first dimension of the LRA is significant ($p < 0.001$), since the observed first part of variance (eigenvalue) is higher than all the other ones in the analyses of the randomly permuted data. If now the coordinates on the first dimension of the LRA are correlated with those of the first dimension of each of the other analyses performed in this chapter, the correlations obtained are given in Table 9.5 below. Notice that a regular correlation on a single dimension is the one-dimensional version of the Procrustes correlation.

Table 9.5 Correlations with first dimension of the LRA of the `RomanCups` data (first dimension of Fig. 9.2, left hand side), with the first dimensions of several alternative analyses in this chapter.

Analysis	Correlation	% variance on 1st dimn
(Fig. 9.2, RHS) 10 logratios	0.963	68.8 %
(Fig. 9.3, LHS) 6 parts	0.998	71.1 %
(Fig. 9.3, RHS) 5 selected logratios	0.963	70.3 %
(Fig. 9.4) 10 selected amalgamation logratios	0.980	90.8 %

Of all the analyses that involve selected logratios, it is the last one that has its first dimension most correlated (0.980) with that of the LRA. The amalgamation ratio *Si/(Na+Ca)* is the clear driver of the first dimension. In fact, the correlation between this ratio and the first dimension of the original LRA of the full data set is computed as 0.977. With its 64.5 % explanation of the logratio variance, log(*Si/(Na+Ca)*) might well be the single best summary of the data, reducing the complete compositional data set to just one logratio involving three parts.

Finally, notice that at some steps there are logratios competing for entering the solution, explaining the same percentage of variance, due to their forming cycles of dependence. To break the tie, the logratio can be chosen at random from the set of competitors, or by optimizing the Procrustes fit (if a statistical criterion is required), or according to substantive reasons (if a decision by the practitioner is available). Further examples are given in Appendix C using the function `STEP` in the `easyCODA` package. See also the Bibliography (Appendix B).

Summary: Simplifying the task: variable selection

- Assuming that the total logratio variance is the measure of variability in a compositional data set, a search can be made amongst the logratios themselves to select those that best explain the total variance. A geometrically-based approach is to select logratios that have a structure similar to that of the complete set of logratios, where this similarity is measured by the Procrustes correlation.
- For a J-part composition, its total logratio variance can be fully explained by $J - 1$ linearly independent logratios. These logratios will necessarily form an acyclic connected graph.
- Every set of ALRs is an acyclic graph, but some ALRs are better than others in terms of Procrustes correlation with the logratio geometry.
- A stepwise search amongst all the logratios can provide a set even smaller than $J - 1$ that explains almost all of the variance and has a structure very similar to the original one. This simplifies the results and interpretation for the practitioner by reducing attention to a parsimonious set of logratios.
- This parsimonious set of logratios contains a reduced set of compositional parts, which can form a subcomposition that explains the essential data structure.

Chapter 10
Case study:
Fatty acids of marine amphipods

This chapter describes, in the form of a short case study, the analysis of a compositional data set that has all the aspects studied in this book. This is the same FattyAcids data set that has been described in Chapters 7 and 8 in the context of amalgamations and zero replacement respectively. Here the extended data set will be treated, where there are several additional variables that serve as predictors of, or responses to, the compositional data. A comprehensive analysis of these data is undertaken to show the power of compositional data analysis in a real-life application. The context here is marine biology, but the way these data are tackled can be carried over in an analogous manner to any other field of research, such as geochemistry, genetics or social science.

10.1 Introduction

Marine amphipods (e.g. Fig. 10.1) are small crustaceans that are a key part of the Arctic marine food chain. They typically feed on herbivorous zooplankton and transfer energy in the form of lipids (fatty acids) to fish species that feed on the amphipods. They are also able to biosynthesize fatty acids into more complex ones, which is an important process for the whole marine ecosystem, right up to mammals such as seals and whales. Hence, studying their fatty acid composition is an important aspect of marine biological research towards the understanding of the so-called *food web*. The data set studied here has fatty acid data from four different

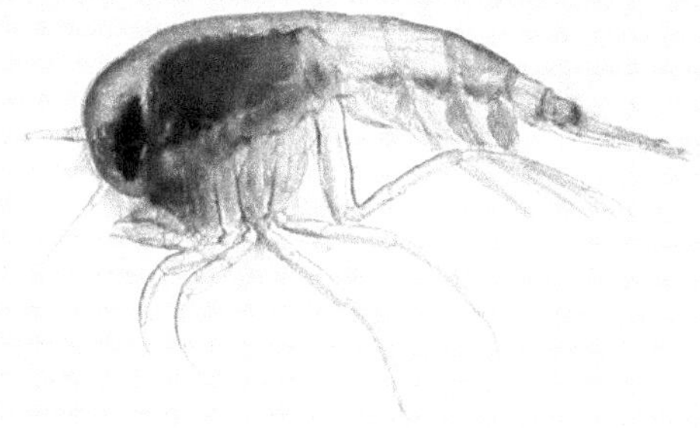

Fig. 10.1 Example of an amphipod of the *Themisto* genus. Body sizes can reach lengths of 15-30 mm. *Source*: neba.arcticresponsetechnology.org.

amphipod species, obtained in two different seasons, summer and winter. The objective is also to compare the compositions of the different species and between the seasons.

10.2 Material and methods

The `FattyAcids` data set contains estimated fatty acid compositions from $n = 52$ individuals, distributed over the four species *Themisto abyssorum*, *T. compressa*, *T. libellula* and *Cyclocaris guilelmi* (notice that the genus name *Themisto* is abbreviated as *T.* after the first occurrence). No explanation is given here of the way these compositions are measured, but see the Appendix B for a reference. For each of the species there are samples from the summer period and from the winter period, as follows:

Species	Summer	Winter	Total
T. abyssorum	6	3	9
T. compressa	6	6	12
T. libellula	8	2	10
C. guilelmi	15	6	21

There are 27 fatty acids (abbreviated as FAs from now on) in the data set: five saturated (SFA), 11 monounsaturated (MUFA) and 11 polyunsaturated (PUFA):[1]

FA category	List
Saturated	*14:0, 15:0, 16:0, 17:0, 18:0*
Monounsaturated	*16:1(n-7), 16:1(n-5), 18:1(n-9), 18:1(n-7), 20:1(n-11), 20:1(n-9), 20:1(n-7), 22:1(n-11), 22:1(n-9), 22:1(n-7), 24:1(n-9)*
Polyunsaturated	*16:2(n-4), 16:3(n-4), 16:4(n-1), 18:2(n-6), 18:3(n-3), 18:4(n-3), 20:4(n-6), 20:4(n-3), 20:5(n-3),22:5(n-3), 22:6(n-3)*

In order to compute logratios, the zeros in the data set are replaced with half the lowest positive value for each FA that has zeros, as described in Chap. 8. Two logratio analyses are performed, one of the 52 samples without taking into account their species and seasonal classifications, and the other focused on discriminating between the eight species–season groups. The latter method is an LRA where the logratios within each group have been averaged, and it is these eight multivariate log-ratio means that are analysed, weighting each mean vector by the group size. This is called a centroid discriminant analysis (see Appendix B, Web Resources) and the individual samples are then added to this centroid discriminant analysis as supplementary points. In both of these analyses, 95 % confidence ellipses will be

---

[1] FAs are long chains of hydrocarbons ending starting with a carboxy group ($-$COOH) and ending with a methyl group (CH3). They are denoted by XX:Y(n-Z), where XX = total number of carbons, Y = number of double bonds of carbon, and Z = position of the double bond closest to the methyl (or omega) end. Saturated FAs have no double bonds, e.g. *14:0*; monounsaturated FAs have one bond, e.g. *16:1(n-7)*; polyunsaturated FAs have two or more double bonds, e.g. *16:2(n-4), 18:4(n-3)*.

drawn around the logratio means of the eight species–season groups, to show how well the groups are separated statistically in the two-dimensional displays. Additionally, in both analyses the separation of the species–season groups is measured by the ratio of between-group logratio variance to total logratio variance.

In the centroid discriminant analysis a stepwise logratio selection is performed to find a small subset of logratios that separates the species–season groups. Finally, a permutation test is conducted to test whether the species–season groups are significantly different in their FA compositions, and if so, whether the difference is between species, between seasons, or a combination (i.e. interaction) of both.

10.3 Results

LRA of the individual samples

Here the information about the species and the seasons is not used in the analysis, but is used afterwards to annotate the plot, showing the means of the eight groups and their dispersions — see Fig. 10.2. The first dimension separates the winter samples on the left from the summer samples on the right, while the second dimension separates the *C. guilelmi* samples at the top from the *Themisto* samples below. The confidence ellipses suggest that the winter–summer compositional differences for all three *Themisto* species are statistically different, and that *C. guilelmi* is significantly different from the *Themisto* species, especially *T. libellula*.

Only the highly contributing species are labelled. It appears that the ratio *18:1(n-9)/18:4(n-3)* lines up with the summer–winter contrast, and that *20:4(n-3)/22:1(n-11)* separates out the winter samples, while *16:0/20:5(n-3)* (for example) separates out the summer samples.

The total (weighted) logratio variance is 0.4525, a much higher value than in other data sets studied in this book — this exemplifies the much higher variability in this type of biochemical data. The between-group variance, i.e. that part of the total variance due to between-group differences, is 0.1906, which is 42.1% of the total. Restricted to the reduced two-dimensional plot of Fig. 10.2, the between-group variance is reduced to 0.1369. Thus in Fig. 10.2, relative to the variance in two dimensions, 71.9% is between-group variance (0.1369 out of 0.1906), which is 30.3% of the total (0.1369 out of 0.4525).

Centroid discriminant LRA between groups

The analysis shown in Fig. 10.2 places an equal weight of 1/52 on each of the 52 individual samples. The group means, or *centroids*, get no weight and are shown *a posteriori* as supplementary points and connected to their respective sample points. Now, in the logratio transformed matrix, the weight is taken off the individual samples, which now all have zero weights, and placed on the group means, in proportion to their respective sample sizes. For example, the group T.aS includes 6 of the 52 samples, and so gets a weight of 6/52. The analysis now treats the individual

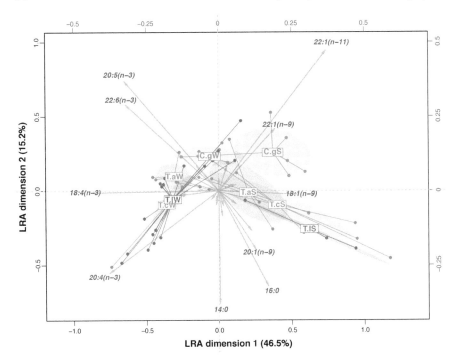

Fig. 10.2 LRA contribution biplot of the FattyAcids data set. T.a = *T. abyssorum*, T.c = *T. compressa*, T.l = *T. libellula*, C.g = *C. guilelmi*; S = Summer, W = Winter. Confidence ellipses (95 %) are shown for the group means (means are shown by labelled boxes). The sample points (dots) are joined to their respective means (group T.cW, lower left, has only two points, so no ellipse is possible). Only the fatty acids highly contributing to the two-dimensional solution are labelled, corresponding to the long arrows from the centre.

points as supplementary and the display of the centroids is optimized. Hence, a better separation of the groups is expected, both visually and in terms of between-group variance. Fig. 10.3 shows the result.

Indeed, in Fig. 10.3, the between-group variance is now increased to 0.1606 (compared to 0.1369 in Fig. 10.2). Hence, 84.4 % of the displayed variance is now between-group (0.1606 out of 0.1906, the same as the sum of the percentages on the two dimensions), and that is 35.5 % of the total (0.1606 out of 0.4525).

In the contribution biplot of Fig. 10.3, notice how much less within-group variance there is for groups such as T.aS and T.lW, although the within-group variance of C.gW has actually increased. It also seems that the gain in between-group separation has been more for the summer samples than the winter ones, with a clearer separation between groups C.gS, T.lS and the other two summer groups that have overlapping confidence ellipses..

The set of highly contributing FAs has also changed, and some have changed direction; for example, *20:1(n-9)* was oriented towards lower right in Fig. 10.2, but is now pointing vertically upwards (but notice that that the position of the group T.lS

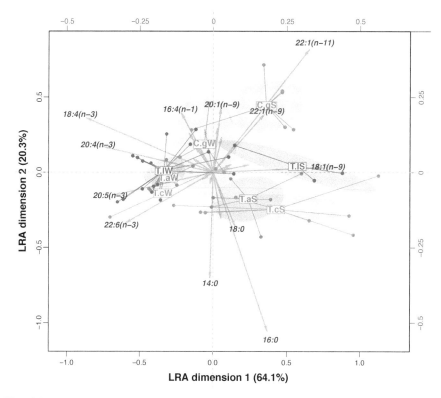

Fig. 10.3 LRA contribution biplot of the FattyAcids data set, where the objective is to separate the species–season groups, as opposed to separating the individual samples. Labels, points and ellipse descriptions as for Fig. 10.2.

has also moved upwards). The main drivers appear to be, for example, *22:1(n-11)*, *18:4(n-3)*, *14:0)* and *16:0*, but it is the ratios that should be investigated for explaining the group differences.

Logratio selection to discriminate between groups

To obtain a small set of logratios that explain the group differences, a stepwise search as described in Chap. 9 can be undertaken[2].

Notice that the procedure is slightly different from that described in Chap. 9, where the objective was to explain variance between individual samples. Here, the aim is not to explain the total logratio variance 0.4525, but rather to explain the between-group variance of 0.1906. Now the eight species–season groups occupy a subspace of only seven dimensions, so a maximum of seven independent logratios is

---

[2] Ideally, this should be done in collaboration with a biochemist who is familiar with amphipods, since at each step there are competing ratios for entering the list. In Appendix B, a reference is given to an article where exactly that strategy is followed in order to obtain a set of logratios that have good statistical properties as well as a clear biochemical interpretation.

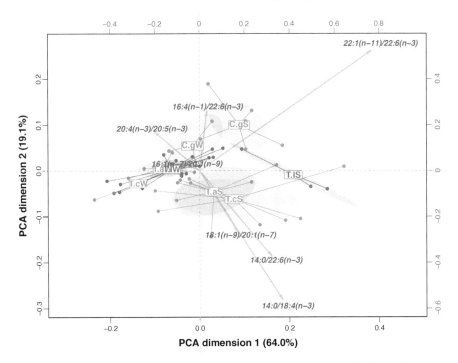

Fig. 10.4 PCA contribution biplot of the `FattyAcids` data set, separating the the species–season groups, using a reduced set of seven FA ratios (Table 10.1) that explain 100 % of the between-group variance. Labels, points and ellipse descriptions as for Fig. 10.2.

required to explain all the between-group variance. A set of seven logratios was chosen stepwise to maximize the Procrustes agreement with the configuration of group centroids (see Table 10.1). The PCA of these logratios is given in Fig. 10.4, and there is a strong agreement with Fig. 10.3. It should be remembered that Fig. 10.3 is based on the full data set of 351 logratios, whereas Fig. 10.4 uses just seven.

Permutation tests for group differences

The `vegan` package also includes permutation testing of the multivariate differences between groups. Suppose that the between-season differences are to be tested against the null hypothesis of no difference. The part of variance explained by the seasonal difference is computed, and is equal to 25.1 % of the total variance. Under the null hypothesis, the labels "Summer" and "Winter" can be randomly re-assigned to the samples, and the variance explained after this random assignment computed. This is repeated 999 times to obtain a null distribution of the part of explained variance, generated 999 times, plus the actual observed value of 25.1 %. It turns out that none of the randomly generated percentages exceeds the observed value, hence the p-value for the test is $p < 0.001$.

After including the seasonal effect, the species effect can be similarly tested, and is also significant at $p = 0.003$, which means that the observed value was the third in the total list of 1000 values in the simulated permutation distribution (including the observed value). Finally, the interaction effect can be added, but this turns out to be nonsignificant, with an estimated p-value of $p = 0.29$. These p-values are all estimates, since they depend on the random permutations, but can be made more accurate by increasing the number of permutations.

Comparison with correspondence analysis

As illustrated in Chap. 8, correspondence analysis (CA) is an alternative that does not need zeros replaced, and generally has low subcompositional incoherence, especially if combined with a power transformation. Fig. 10.5 shows the CA of this amphipod data set, with zeros retained, in the same output style as the previous figures in this chapter. The square-root transformation has been used, which was found in Sect. 8.4 to be close to the optimal transformation both for reducing subcompositional incoherence and for increasing Procrustes correlation. This result is the centroid discriminant version of CA comparable to Fig. 10.3 where the objective is to separate the species–season groups as much as possible. The CA contribution biplot in Fig. 10.5 strongly resembles the LRA one in Fig. 10.3.

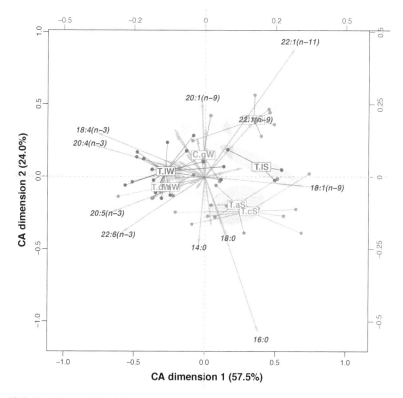

Fig. 10.5 Contribution CA biplot of the `FattyAcids` data set, including original zeros, where the objective is to separate the species–season groups, as opposed to separating the individual samples. The square-root transformation is applied, as described in Sect. 8.4. Labels, points and ellipse descriptions as for Fig. 10.2.

Univariate statistics and comparisons

One of the most useful by-products of the logratio selection approach is to arrive at a few ratios that can be validly reported using regular statistical summaries. For example, the seven ratios used in Fig. 10.4 can be summarized in their untransformed form by their medians and reference ranges — see Table 10.1 (cf. Table 9.3 where this style of table was first used). These ratios could also be used to make univariate comparisons between the species and between the seasons, always remembering that their distributions are highly skewed, hence the preference for using nonparametric or permutation tests.

Table 10.1 Univariate statistics — medians and reference ranges — for the seven selected logratios that separate the species–season groups.

Ratio	Median	Reference range
22:1(n-11) / 22:6(n-3)	0.612	0.075 – 8.925
14:0 / 18:4(n-3)	1.442	0.416 – 72.84
14:0 / 22:6(n-3)	0.668	0.111 – 8.498
20:4(n-3) / 20:5(n-3)	0.065	0.015 – 2.265
16:1(n-7) / 20:1(n-9)	0.539	0.200 – 1.458
16:4(n-1) / 22:6(n-3)	0.051	0.001 – 0.455
18:1(n-9) / 20:1(n-7)	12.52	4.470 – 87.98

10.4 Discussion and conclusion

Researchers who collect compositional data face the same problems as when analysing regular multivariate statistical data. They want to (i) understand relationships between the variables, (ii) make groupings of the samples, (iii) compare pre-defined groups of samples, (iv) identify the most important variables that account for the differences between individual samples as well as between groups of samples, and (v) interpret the results in a way that is understandable and appropriate to the context of the study. The nature of compositional data, specifically the unit-sum constraint, precludes the use of regular statistical methods on the compositional parts. The logratio transformation takes compositional data into a sample space where regular statistical methodology can operate. In this transformed space, logratios can be selected which are the most important in separating individual samples, i.e. the various marine amphipods in the present chapter's application. In addition, the logratios, in this case involving ratios of fatty acids, can be identified which account for the differences between samples grouped by species and by seasons. Whatever the objective, appropriately chosen logratios can effectively replace the original compositional data set, resulting in a major simplification of the data, reducing a complex mix of multivariate relationships, with all its inherent randomness and measurement error, to its essential and parsimonious non-random structure.

Appendix A
Theory of compositional data analysis

Everything should be as simple as possible, but no simpler   *– Albert Einstein*

In this Appendix the mathematical theory of compositional data analysis is summarized. In most cases matrix–vector algebra will be used, and the scalar versions will be given when necessary for clarification. No proofs are given of the results, and readers should refer to the bibliography in Appendix B for more detailed sources on theoretical aspects.

A.1 Basic notation

Compositional data matrix

Compositional data constitute a matrix of nonnegative numbers, with I rows and J columns, denoted by $\mathbf{X}$ $(I \times J)$. By convention, the rows are the I samples or units of observation, $i = 1, 2, \ldots, I$, and the columns the J compositional parts, $j = 1, 2, \ldots, J$. By definition, the compositions in the rows of $\mathbf{X}$ are closed (i.e. add up to 1): $\sum_j x_{ij} = 1$, for all i. Compositional data can include zeros, but in most of what follows from Sect. A.2 onwards, the values in $\mathbf{X}$ must be strictly positive.

Marginal values and weights

The column means are denoted by $c_j, j = 1, 2, \ldots, J$, collected into the vector $\mathbf{c}$ $(J \times 1)$. If $\mathbf{1}$ $(I \times 1)$ denotes the vector of I ones, $\mathbf{c}^\mathsf{T} = (1/I)\mathbf{1}^\mathsf{T}\mathbf{X}$, where $^\mathsf{T}$ denotes the transpose of a vector or a matrix — notice that the vector product $\mathbf{1}^\mathsf{T}\mathbf{X}$ sums all the columns into a row vector. Often the elements of $\mathbf{c}$ will be used as weights for the columns (parts), otherwise different weights can be specified by the user, or the weights can be equal, i.e. $1/J, j = 1, 2, \ldots J$, in which case the analysis is referred to as "unweighted". In what follows the notation $c_j, j = 1, 2, \ldots, J, \sum_j c_j = 1$, will also be used for general positive weights for the columns, where it is understood that they are by default the column means unless otherwise specified. The rows are usually equally weighted by values $1/I, i = 1, 2, \ldots, I$, but can also be redefined according to the context of the data. Row weights are denoted by $r_i, i = 1, 2, \ldots, I$ and collected in a vector $\mathbf{r}$ $(I \times 1)$, where $\sum_i r_i = 1$. For generally weighted rows, the column means are weighted averages: $\mathbf{c}^\mathsf{T} = \mathbf{r}^\mathsf{T}\mathbf{X}$.

A.2 Ratios and logratios

As shown in Chap. 2, ratios of parts obey the principle of subcompositional coherence and will be log-transformed for data analysis — Chap. 3 defined several types of logratios (i.e. log-transformed ratios).

Additive logratios

The simplest logratios are the additive logratios (ALRs), where one part is fixed as the denominator for the other $J - 1$ parts as numerators. Because the parts can be permuted, the chosen denominator can always be arranged to be the last part, so the ALRs can be defined as follows:

$$\text{ALRs:}\quad y_{ij} = \log\left(\frac{x_{ij}}{x_{iJ}}\right) = \log(x_{ij}) - \log(x_{iJ}), \ j = 1, 2, \ldots, J-1 \qquad (A.1)$$

The transformation from ratios to ALRs can be written in matrix format:

$$\mathbf{Y} = \log(\mathbf{X})\,\mathbf{A} \qquad (A.2)$$

where $\log(\mathbf{X})$ is the $I \times J$ matrix of log-transformed data, and $\mathbf{A}$ the $J \times (J-1)$ matrix

$$\mathbf{A} = \begin{bmatrix} 1 & 0 & 0 & \cdots & 0 \\ 0 & 1 & 0 & \cdots & 0 \\ 0 & 0 & 1 & \cdots & 0 \\ \vdots & \vdots & \vdots & \ddots & \vdots \\ 0 & 0 & 0 & \cdots & 1 \\ -1 & -1 & -1 & \cdots & -1 \end{bmatrix} \qquad (A.3)$$

(notice that the matrix in (3.2) is transposed compared to the one above, because there the composition was a column vector, whereas here the compositions are in the rows — this also explains why the multiplication in (A.2) is on the right).

Logratios are weighted by the product of their part weights, i.e. the (j, j')-th logratio has weight $c_j c_{j'}$, which would be $1/J^2$ for an unweighted analysis. Hence, the weights assigned to the $J - 1$ ALRs (columns of $\mathbf{Y}$ in (A.2)) are $c_j c_J, j = 1, 2, \ldots, J-1$, i.e. proportional to the weight of the part in the numerator.

Given the ALRs y_{ij}, the inverse transformation to recover the compositional parts x_{ij} is:

$$\text{inverse ALRs:}\quad x_{ij} = \frac{e^{y_{ij}}}{1 + \sum_{k=1}^{J-1} e^{y_{ik}}}, \ j = 1, 2, \ldots, J-1; \ x_{iJ} = 1 - \sum_{k=1}^{J-1} x_{ik} \quad (A.4)$$

Centred logratios

Centred logratios (CLRs) of the rows are formed by centring the log-transformed parts for each row by its mean, weighted or unweighted. Thus, CLRS are log-transformations of the ratio of the parts to their (weighted) geometric mean:

$$\text{CLRs:}\quad y_{ij} = \log\left(x_{ij} / \prod_{j=1}^{J} x_{ij}^{c_j}\right) = \log(x_{ij}) - \sum_{j=1}^{J} c_j \log(x_{ij}), \ j = 1, 2, \ldots, J \quad (A.5)$$

where the weights c_j are by default the compositional part means, or pre-defined, or all equal to $1/J$ for an unweighted analysis. Again, this is a linear transformation of the log-transformed compositions in the rows of $\log(\mathbf{X})$:

$$\mathbf{Y} = \log(\mathbf{X})\,\mathbf{B} \tag{A.6}$$

where the matrix $\mathbf{B}$ $(J \times J)$ is

$$\mathbf{B} = \begin{bmatrix} 1-c_1 & -c_1 & -c_1 & \cdots & -c_1 \\ -c_2 & 1-c_2 & -c_2 & \cdots & -c_2 \\ \vdots & \vdots & \vdots & \ddots & \vdots \\ -c_J & -c_J & -c_J & \cdots & 1-c_J \end{bmatrix} = \mathbf{I} - \mathbf{c}\mathbf{1}^{\mathsf{T}} \tag{A.7}$$

where $\mathbf{I}$ is the $J \times J$ identity matrix and $\mathbf{1}$ is the $J \times 1$ vector of 1s. The CLRs in the columns of the matrix $\mathbf{Y}$ are weighted by the elements in $\mathbf{c}$.

Given the CLRs y_{ij}, the inverse transformation to recover the compositional parts x_{ij} is:

$$\text{inverse CLRs:} \quad x_{ij} = \frac{e^{y_{ij}}}{1 + \sum_{k=1}^{J} e^{y_{ik}}}, \quad j = 1, 2, \ldots, J \tag{A.8}$$

Isometric logratios

Isometric logratios (ILRs) are a specific set of log-transformed ratios of geometric means, or equivalently differences in (weighted) means of two subsets of log-transformed compositional data. For example, suppose J_1 and J_2 denote disjoint subsets of parts, with the sums of weights in the two subsets, respectively $C_1 = \sum_{j \in J_1} c_j$ and $C_2 = \sum_{j \in J_2} c_j$. Then the corresponding ILR is

$$\text{ILR:} \quad \sqrt{\frac{C_1 C_2}{C_1 + C_2}} \log \frac{\left(\prod_{j \in J_1} x_{ij}^{c_j} \right)^{1/C_1}}{\left(\prod_{j \in J_2} x_{ij}^{c_j} \right)^{1/C_2}}$$

$$= \sqrt{\frac{C_1 C_2}{C_1 + C_2}} \left(\frac{1}{C_1} \sum_{j \in J_1} c_j \log(x_{ij}) - \frac{1}{C_2} \sum_{j \in J_2} c_j \log(x_{ij}) \right) \tag{A.9}$$

The special case of unweighted parts is given in Chap. 3, Eqn. (3.5), but notice that the definition in (A.9), for $c_j = 1/J$, $j = 1, \ldots, J$, which is the form of the unweighted version given in this book, gives the result of (3.5) divided by $\sqrt{J}$.

One way to choose a set of ILRs is to use the groupings based on a hierarchical clustering of the parts, where each of the $J - 1$ nodes of the cluster dendrogram defines a contrast between two subsets of parts — see Sect. 7.4 for an example using unweighted ILRs. Once a complete set of ILRs is fixed, they can also be expressed as a matrix transformation (not shown here), and can be inverted as a group back to the original parts in another matrix operation — see the simpler special case of pivot logratios next.

Pivot logratios

Pivot logratios (PLRs) are ILRs where each part is contrasted against the remainder in an ordered list: the first part versus the rest, then the second part versus the rest (but excluding the first), the third part versus the rest (but excluding the first and second), and so on, giving $J - 1$ ratios where the numerators are single parts $j = 1, 2, \ldots, J - 1$. Here again the general weighted case is defined:

$$\text{PLRs:} \quad \sqrt{\frac{c_k C_{J-k}}{c_k + C_{J-k}}} \log \frac{X_k}{[\prod_{j=k+1}^{J} X_j]^{c_j/C_{J-k}}}$$

$$= \sqrt{\frac{c_k C_{J-k}}{c_k + \cdots + c_J}} \left(\log(X_k) - \frac{1}{C_{J-k}} \sum_{j=k+1}^{J} c_j \log(X_j) \right), \quad k = 1, \ldots, J-1.$$

$$(A.10)$$

where $C_{J-k} = c_{k+1} + c_{k+2} + \cdots + c_J$, the combined weights of the last $J - k$ parts in the ordered list. The special case of unweighted parts is given in Chap. 3, Eqn. (3.7) — again, the definition here in (A.10) for equal weights $c_j = 1/J$, $j = 1, \ldots, J$, will give the value according to definition (3.7) divided by $\sqrt{J}$.

The matrix transformation for obtaining PLRs is $\mathbf{Y} = \log(\mathbf{X}) \, \mathbf{C}$, where

$$\mathbf{C} = \begin{bmatrix} 1 & 0 & 0 & \cdots & 0 \\ -c_2/C_{J-1} & 1 & 0 & \cdots & 0 \\ -c_3/C_{J-1} & -c_3/C_{J-2} & 1 & \cdots & 0 \\ \vdots & \vdots & \vdots & \ddots & \vdots \\ -c_{J-1}/C_{J-1} & -c_{J-1}/C_{J-2} & -c_{J-1}/C_{J-3} & \cdots & 1 \\ -c_J/C_{J-1} & -c_J/C_{J-2} & -c_J/C_{J-3} & \cdots & -1 \end{bmatrix} \mathbf{D} \qquad (A.11)$$

where $\mathbf{D}$ is the diagonal matrix of normalizing constants:

$$\sqrt{(c_1 C_{J-1})}, \sqrt{(c_2 C_{J-2})/(c_2 + \cdots + c_J)}, \sqrt{(c_3 C_{J-3})/(c_3 + \cdots + c_J)}, \ldots, \sqrt{c_{J-1}/c_J} \, .$$

The inverse operation, to return from the set of PLRs in $\mathbf{Y}$ to the compositional parts is then, as for all ILR balances, in matrix form:

$$\text{inverse PLRs:} \quad \tilde{\mathbf{X}} = \exp(\mathbf{Y}) \, \mathbf{C}^{\mathsf{T}} \text{ then close rows of } \tilde{\mathbf{X}} \text{ to obtain } \mathbf{X} \qquad (A.12)$$

where $\exp(\mathbf{Y})$ is the matrix of exponentiated values $\exp(y_{ij}) = e^{y_{ij}}$.

Amalgamation logratios

Logratios of amalgamations are simpler constructs than ILRs and PLRs. Amalgamations are generally constructed according to the substantive nature of the study rather than statistical criteria. They are then expressed as ratios with respect to single parts or other amalgamations to give amalgamation logratios, denoted by SLRs (i.e. "summed" logratios) to avoid confusion with ALRs (additive logratios). A set of amalgamation balances can also be constructed according to a recursive partitioning of the parts suggested by a cluster analysis or any substantive argument.

A.3 Logratio distance

The concept of distance underlies all methods of multivariate analysis of a compositional data set. Logratio distances can be defined in a symmetric way between samples (rows) and between parts (columns). The formula for the logratio distance between rows in its most general form, including weights, is as follows, between rows i and i' of $\mathbf{X}$:

$$d_{ii'} = \sqrt{\sum\sum_{j<j'} c_j c_{j'} \left[\log\frac{x_{ij}}{x_{ij'}} - \log\frac{x_{i'j}}{x_{i'j'}}\right]^2} = \sqrt{\sum\sum_{j<j'} c_j c_{j'} \left[\log\frac{x_{ij}x_{i'j'}}{x_{ij'}x_{i'j}}\right]^2}$$

(A.13)

where $\sum\sum_{j<j'}$ denotes summation over the unique $J(J-1)/2$ pairwise logratios. The special case of unweighted parts, i.e. $c_j = 1/J$, is given in Eqn. (2.3).

The distance between columns (parts) is the same definition, simply exchanging the indices of the summation and the weights, which are now row weights r_i, usually equal: $r_i = 1/I$:

$$d_{jj'} = \sqrt{\sum\sum_{i<i'} r_i r_{i'} \left[\log\frac{x_{ij}}{x_{i'j}} - \log\frac{x_{ij'}}{x_{i'j'}}\right]^2} = \sqrt{\sum\sum_{i<i'} r_i r_{i'} \left[\log\frac{x_{ij}x_{i'j'}}{x_{i'j}x_{ij'}}\right]^2}$$

(A.14)

Notice that in the definitions of both inter-row and inter-column distances (right hand sides of Eqns. (A.13) and (A.14)), the sums are over the same logarithms of cross-products $\log[(x_{ij}x_{i'j'})/(x_{ij'}x_{i'j})]$, first summed over pairs of columns and then over pairs of rows, respectively.

The same inter-row distances can be computed more efficiently by summing just over the columns of the CLRs, which are row-centred. That is, for the row CLRs y_{ij} defined in (A.4), the inter-row logratio distance (A.13) is the same as

$$d_{ii'} = \sqrt{\sum_j c_j (y_{ij} - y_{i'j})^2}$$

(A.15)

The corresponding simpler form of the inter-column distances uses the CLRs computed columnwise, i.e. column-centred by subtracting the column means from the columns: $z_{ij} = \log(x_{ij}) - \sum_i r_i \log(x_{ij})$, and then applying the corresponding version of Eqn. (A.15) where summation is over the rows:

$$d_{jj'} = \sqrt{\sum_i r_i (z_{ij} - z_{ij'})^2}$$

(A.16)

A.4 Logratio variance

The total variance in a data set is an important measure of the variability between samples, and forms the target measure when explaining variance by some additional predictors or by some well-chosen logratios in the same data set.

For univariate data, variance is the weighted average sum-of-squared differences between the data values and their mean. If the I sample values x_i have positive weights r_i, $i = 1, 2, \ldots, I$, where $\sum_i r_i = 1$, then their weighted mean, or centroid, is $\bar{x} = \sum_i r_i x_i$. The unweighted case is when $r_i = 1/I$ for all i. The weighted variance, sometimes called inertia, is the weighted average of the differences between the values and their centroid: $\sum_i r_i (x_i - \bar{x})^2$. This quantity is identical to the weighted sum-of-squared differences between all pairs of observations: $\sum \sum_{i<i'} r_i r_{i'} (x_i - x_{i'})^2$, where the (i, i')-th pair has weight equal to the product of the weights $r_i r_{i'}$.

For multivariate data such as compositional data, the total logratio variance is a simple extension of the univariate case, averaging the variances over all pairwise logratios, or equivalently over all CLRs. For example, each of the $J(J-1)/2$ pairwise logratios has a mean and a variance[1] $\text{var}_{jj'}$, and the total variance is then the weighted average of these variances, where the weight of the (j, j')-th logratio is $c_j c_{j'}$, i.e. $\sum \sum_{j<j'} c_j c_{j'} \text{var}_{jj'}$. Equivalently, if the variance of the j-th CLR is var_j, then the total logratio variance is the weighted average $\sum_j c_j \text{var}_j$.

As in the univariate case, the total variance is identical to the weighted sum-of-squared distances between samples (notice here that multivariate 'distance' replaces univariate 'difference'): $\sum \sum_{i<i'} r_i r_{i'} d_{ii'}^2$, where $d_{ii'}$ is defined in Eqns. (A.13) and (A.15). Similarly, there is a symmetric formulation of the total logratio variance as $\sum \sum_{j<j'} c_j c_{j'} d_{jj'}^2$, where $d_{jj'}$ is defined in Eqns. (A.14) and (A.16).

Although it is not the most efficient way of computing the total variance, the following formula shows the symmetry between rows and columns in its formulation:

$$\text{total variance} = \sum \sum_{i<i'} \sum \sum_{j<j'} r_i r_{i'} c_j c_{j'} \left[\log \frac{x_{ij} x_{i'j'}}{x_{ij'} x_{i'j}} \right]^2 \tag{A.17}$$

A.5 Logratio analysis (LRA)

A J-part compositional data set has rank less than or equal to $J-1$, and both the samples (rows) and compositional parts (columns) occupy $(J-1)$-dimensional spaces. Logratio analysis (LRA) aims to find fewer dimensions that retain a maximum amount of the total logratio variance, with the aim of separating non-random variation in a few "principal" dimensions from random variation in the others. The analysis is an application of the *singular value decomposition* (SVD), one of the most useful results in matrix algebra. The SVD is the decomposition of a matrix into the product of three matrices of simple structure $\mathbf{S} = \mathbf{U} \mathbf{D}_\alpha \mathbf{V}^\mathsf{T}$, where the *left* and *right* *singular vectors* in the columns of $\mathbf{U}$ and $\mathbf{V}$, respectively, satisfy the orthonormality conditions: $\mathbf{U}^\mathsf{T} \mathbf{U} = \mathbf{V}^\mathsf{T} \mathbf{V} = \mathbf{I}$, and $\mathbf{D}_\alpha$ is the diagonal matrix of positive singular values in descending order: $\alpha_1 \geq \alpha_2 \geq \cdots > 0$.

Unless specified otherwise, LRA will be weighted, including weights c_j for the compositional parts (columns) and weights r_i for the samples (rows). The usual default is to equally weight the rows but weight the columns by their mean values.

---

[1] The usual computation of the sample variance in most software packages, as well as in R using function `var`, divides by the sample size minus 1, i.e. $I-1$ in the present notation, whereas in the unweighted case presented here it is preferred to divide by the sample size I, giving each sample point an equal weight of $1/I$. For weighted variances in R, use the function `cov.wt`, for example.

The steps for performing an LRA are as follows, starting from the matrix of log-transformed compositional data, $\log(\mathbf{X})$:

LRA1. Double-centre[2] the matrix $\log(\mathbf{X})$: $\quad \widetilde{\mathbf{Y}} = (\mathbf{I} - \mathbf{1r}^{\mathsf{T}})\log(\mathbf{X})(\mathbf{I} - \mathbf{1c}^{\mathsf{T}})^{\mathsf{T}}$

$$(A.18)$$

LRA2. Apply weights to rows and columns: $\quad \mathbf{S} = \mathbf{D}_r^{\frac{1}{2}} \widetilde{\mathbf{Y}} \mathbf{D}_c^{\frac{1}{2}} \qquad (A.19)$

LRA3. Perform SVD: $\quad \mathbf{S} = \mathbf{U}\mathbf{D}_\alpha\mathbf{V}^{\mathsf{T}} \qquad (A.20)$

LRA4. Principal coordinates of rows: $\quad \mathbf{F} = \mathbf{D}_r^{-\frac{1}{2}}\mathbf{U}\mathbf{D}_\alpha \qquad (A.21)$

LRA5. Principal axes: $\quad \mathbf{A} = \mathbf{D}_c^{\frac{1}{2}}\mathbf{V} \qquad (A.22)$

LRA6. Standard coordinates of columns: $\quad \mathbf{\Gamma} = \mathbf{D}_c^{-\frac{1}{2}}\mathbf{V} \qquad (A.23)$

LRA7. Contribution coordinates of columns: $\quad \mathbf{\Gamma}^* = \mathbf{D}_c^{\frac{1}{2}}\mathbf{\Gamma} = \mathbf{V} \qquad (A.24)$

Two-dimensional biplots of the results are given by plotting the row principal coordinates in the first two columns of $\mathbf{F}$ with either the corresponding column standard coordinates in the first two columns of $\mathbf{\Gamma}$ (*asymmetric biplot*) or those of the contribution coordinates in $\mathbf{\Gamma}^*$ (*contribution biplot*).

The two-dimensional *symmetric map* of the LRA is the plotting of the first two columns of the row principal coordinates (A.21) in $\mathbf{F}$ jointly with those of the column principal coordinates $\mathbf{G}$:

LRA8. Principal coordinates of columns: $\quad \mathbf{G} = \mathbf{D}_c^{-\frac{1}{2}}\mathbf{V}\mathbf{D}_\alpha \qquad (A.25)$

This is not a biplot, strictly speaking, but has the practical advantage that the row and column points are equally scaled along the principal axes, their weighted variances both being equal to the amount of variance explained on the axes, i.e. the eigenvalues or squared singular values α_k^2 on axis k, $k = 1, 2$.

The sum of the squared singular values, $\sum_k \alpha_k^2$ (i.e. sum of eigenvalues), equals the total logratio variance and α_k^2 is the part of variance explained by axis k, usually expressed as a percentage of this total.

A.6 Principal component analysis (PCA)

Principal component analysis (PCA) is another variant of the SVD, which can be used for regular interval-scaled variables. Here it is applied to matrices of logratios derived from a set of compositional data — these can be logratios of any type and of any size.

The first three steps for performing a PCA of any set of logratios are as follows, starting from a matrix $\mathbf{L}$ of J^* logratios, with weights in $\mathbf{w}$ $(J^* \times 1)$, in diagonal

---

[2] Notice that this is equivalent to column-centring the matrix $\mathbf{Y}$ of CLRs, which is already row-centred: $\widetilde{\mathbf{Y}} = (\mathbf{I} - \mathbf{1r}^{\mathsf{T}})\mathbf{Y}$, where $\mathbf{Y} = \log(\mathbf{X})(\mathbf{I} - \mathbf{1c}^{\mathsf{T}})^{\mathsf{T}} = \log(\mathbf{X})(\mathbf{I} - \mathbf{c1}^{\mathsf{T}})$ — see (A.7).

matrix $\mathbf{D}_w$). For example, in the weighted approach, the default weights of a simple logratio formed by parts j and j' would be the products $c_j c_{j'}$ of the weights of the constituent parts; or if the part in the denominator, for example, were an amalgamation J', then the corresponding weight of the logratio would be $c_j c_{J'}$, where $c_{J'}$ is the sum of the weights of the parts in the amalgamation.

PCA1. Column-centre the matrix $\mathbf{L}$: $\widetilde{\mathbf{L}} = (\mathbf{I} - \mathbf{1r}^\mathsf{T})\mathbf{L}$ (A.26)

PCA2. Apply weights to rows and columns: $\mathbf{S} = \mathbf{D}_r^{\frac{1}{2}} \widetilde{\mathbf{L}} \mathbf{D}_w^{\frac{1}{2}}$ (A.27)

PCA3. Perform SVD: $\mathbf{S} = \mathbf{U}\mathbf{D}_\alpha \mathbf{V}^\mathsf{T}$ (A.28)

The remaining steps are identical to (A.21–25) of LRA, using the diagonal matrix $\mathbf{D}_w$ of weights instead of $\mathbf{D}_c$.

The PCA of the full set of CLRs is equivalent to LRA — in this case the weights are the part weights c_j. The PCA of the full set of of $J^* = \frac{1}{2}J(J-1)$ logratios is also equivalent to LRA, the only difference being that in the LRA only the J parts are plotted and the logratios are represented by the links between pairs of parts.

A.7 Procrustes analysis

Procrustes analysis (PRO) is a method for matching two multidimensional configurations by introducing translation, rotation and scaling operations to make them as similar as possible to each other. It is used in compositional data analysis to measure how similar two data structures are, for example between the matrix $\mathbf{F}_1$ of principal coordinates from an LRA and the matrix $\mathbf{F}_2$ of principal coordinates from a PCA of a reduced set of logratios. PRO is another application of the SVD and the way it has been used in this book is defined by the following series of steps. Both $\mathbf{F}_1$ and $\mathbf{F}_2$ are assumed to have already been column-centred, which takes care of the translation operation, since this makes their means identical. The first step is then to normalize the two configurations so they both have sums of squares equal to 1, which takes care of the scaling. It just remains to find the rotation of one configuration to agree as closely as possible with the other, which is where the SVD is used.

PRO1. Normalize both matrices[3]: $\mathbf{F}_1^* = \mathbf{F}_1 / \sqrt{\mathrm{trace}(\mathbf{F}_1^\mathsf{T}\mathbf{F}_1)}$

$\mathbf{F}_2^* = \mathbf{F}_2 / \sqrt{\mathrm{trace}(\mathbf{F}_2^\mathsf{T}\mathbf{F}_2)}$ (A.29)

PRO2. Compute cross-products: $\mathbf{S} = \mathbf{F}_1^{*\mathsf{T}}\mathbf{F}_2^*$ (A.30)

PRO3. Perform SVD: $\mathbf{S} = \mathbf{U}\mathbf{D}_\alpha \mathbf{V}^\mathsf{T}$ (A.31)

PRO4. Rotation matrix: $\mathbf{Q} = \mathbf{V}\mathbf{U}^\mathsf{T}$ (A.32)

---

[3] The notation "trace" denotes the trace of a matrix (the sum of its diagonal values). The diagonal elements of $\mathbf{A}^\mathsf{T}\mathbf{A}$ (or $\mathbf{A}\mathbf{A}^\mathsf{T}$) contain the sums of squares of all the elements in the columns and in the rows, respectively, hence trace($\mathbf{A}^\mathsf{T}\mathbf{A}$) is the sum of squares of all the elements of $\mathbf{A}$.

PRO5. Sum-of-squared errors: $\quad E = \mathrm{trace}[(\mathbf{F}_1^* - \mathbf{F}_2^*\mathbf{Q})^\mathsf{T}(\mathbf{F}_1^* - \mathbf{F}_2^*\mathbf{Q})]$ (A.33)

PRO6. Procrustes correlation: $\quad r = \sqrt{1 - E}$ (A.34)

A.8 Constrained logratio analysis and redundancy analysis

When linear constraints are imposed on the PCA solution, the resultant method is called redundancy analysis (RDA), also called PCA on instrumental variables. Constraints are usually imposed using external continuous or categorical explanatory variables in a samples-by-variables matrix $\mathbf{Z}$, where categorical variables are coded as dummy variables, as in a regression analysis. A constrained version of LRA can be obtained by performing an RDA on the CLR matrix, for example.

RDA consists of an extra intermediary step, where the appropriately transformed data matrix (e.g. the double-centred matrix in (A.18)) is projected onto the space defined by the explanatory variables in $\mathbf{Z}$. For example, between steps LRA1 and LRA2, i.e. (A.18) and (A.19), the following regression-type step is interposed, assuming that the columns of $\mathbf{Z}$ are standardized to have mean 0 and variance 1:

RDA. Project $\widetilde{\mathbf{Y}}$ onto $\mathbf{Z}$: $\qquad\qquad \mathbf{Y}^* = \mathbf{Z}(\mathbf{Z}^\mathsf{T}\mathbf{Z})^{-1}\mathbf{Z}^\mathsf{T}\widetilde{\mathbf{Y}}$ (A.35)

After that the algorithm follows that of LRA from step LRA2 (A.19) onwards, with $\mathbf{Y}^*$ replacing $\widetilde{\mathbf{Y}}$. To accommodate a singular matrix $\mathbf{Z}^\mathsf{T}\mathbf{Z}$ in (A.35), for example when a full set of dummy variables is included, the Moore-Penrose generalized inverse $(\mathbf{Z}^\mathsf{T}\mathbf{Z})^+$ is conveniently used.

The result of the dimension reduction provides coordinates of the samples and compositional parts in the reduced space that is constrained to be directly related to the explanatory variables. The continuous explanatory variables can be shown with respect to the first two dimensions, say, by correlating them with the solution dimensions and using the correlation coefficients as coordinates. In the case of categorical explanatory variables, the categories can be displayed at the average positions of the sample points belonging to the respective categories.

In Chap. 9 RDA is not used explicitly for dimension reduction, but rather to obtain the part of explained variance due to logratios. The RDA step above splits $\widetilde{\mathbf{Y}}$ into two matrices: $\widetilde{\mathbf{Y}} = \mathbf{Y}^* + (\widetilde{\mathbf{Y}} - \mathbf{Y}^*)$, the first is in the space of the explanatory variables $\mathbf{Z}$ and the second in the space that is linearly uncorrelated with the explanatory variables. The total variance, equal to $\mathrm{trace}(\widetilde{\mathbf{Y}}^\mathsf{T}\widetilde{\mathbf{Y}})$, is similarly decomposed into two parts, the variance in the space of $\mathbf{Z}$, equal to $\mathrm{trace}(\mathbf{Y}^{*\mathsf{T}}\mathbf{Y}^*)$ and the residual variance in the space uncorrelated with the explanatory variables, equal to $\mathrm{trace}[(\widetilde{\mathbf{Y}} - \mathbf{Y}^*)^\mathsf{T}(\widetilde{\mathbf{Y}} - \mathbf{Y}^*)]$. The proportion of explained variance is thus $\mathrm{trace}(\mathbf{Y}^{*\mathsf{T}}\mathbf{Y}^*)$ relative to $\mathrm{trace}(\widetilde{\mathbf{Y}}^\mathsf{T}\widetilde{\mathbf{Y}})$.

A.9 Permutation tests

Permutation tests can be conducted to assess the statistical significance of the percentage of explained variance in RDA or of the Procrustes correlation, against a null model of completely random data. The principle is to generate a null distribution of values, assuming no relationship between the response matrix and predictors (for the RDA) or between the two matrices being fitted to each other (for PRO).

For the RDA, suppose the percentage of variance explained by a set of predictors is equal to T_0, which is the value of the statistic to be tested. The rows of the predictor matrix are randomly permuted, and the analysis is repeated to obtain the explained variance, say T_1. Do this a large number of times, 999 times is typical, thus obtaining enough to define a null distribution of the statistic. Now add the observed T_0 to those under random permutations, $T_1, \ldots, T_{999}$, to bring the set up to 1000 values. Count how many of these 1000 values are larger than or equal to T_0 — at least one of them is, the T_0 that was originally observed. The p-value for the test is that count divided by 1000. Hence, if T_0 is the largest of them all, then $p = 0.001$. If 9999 values were generated under random permutations, and T_0 is still the highest, the p-value would be $p = 0.0001$. The number of potential permutations is usually extremely large, so the p-value is always an estimate. But by increasing the number of permutations, and with knowledge of the distribution of a proportion, the p-value can always be estimated exactly with high confidence for a prescribed precision.

For the Procrustes test, a similar strategy is followed. The rows of one of the two matrices are randomly permuted many times, computing the Procrustes correlation each time, and then comparing the correlation for the original unpermuted data with the distribution of those based on the permuted data. The p-value is estimated by the proportion of correlations greater than or equal to the observed one.

A.10 Weighted Ward clustering

Weighted Ward clustering is performed either on the rows or the columns in a completely symmetric way. It is illustrated here for the rows (samples) — in Sect. 7.3 clustering of the columns (compositional parts) was illustrated. The clustering criterion at the start of the process is:

$$\frac{r_i r_{i'}}{r_i + r_{i'}} d_{ii'}^2 \tag{A.36}$$

where $d_{ii'}$ is the logratio distance (A.15) between rows i and i'. The smallest value of (A.36) determines which pair of samples is clustered, forming a node of the clustering tree, or dendrogram.. Supposing that the pair of clustered samples is denoted by I^*, the corresponding rows of the matrix Y of CLRs (see (A.6)) are combined in a weighted average with elements:

$$\tilde{y}_{I^*, j} = (r_i y_{ij} + r_{i'} y_{i'j})/(r_i + r_{i'}) \quad j = 1, \ldots, J \tag{A.37}$$

and the new row replacing rows i and i' is assigned a weight $r_i + r_{i'}$. The distances between this cluster I^* of samples and the other samples is updated and the clustering continues in the same way, combining the closest pairs according to (A.36), recomputing the rows as weighted averages and choosing the minimum value of the criterion at each step of the clustering. The total logratio variance will be decomposed exactly as the sum of the values of (A.36) at the nodes of the dendrogram.

A variation of this algorithm is to update using weighted averages of the original compositions, not CLRs. Similarly, for column clustering, parts are simply combined by amalgamation, producing nodes with a straightforward interpretation.

Appendix B
Bibliography of compositional data analysis

No bibliographic references have been given in the chapters and Appendix A. In this appendix the main references for the different parts of the book are given, with annotations and a guide for further reading, with an accent on the literature sources that are the most useful for applied researchers. The publications are given in time order, and are not intended to be an exhaustive list.

B.1 Books

- John Aitchison (1986) *The Statistical Analysis of Compositional Data.* Chapman & Hall, London. Reprinted 2003 with additional material by Blackburn Press.

This book remains a rich source of the foundations of compositional data analysis according to Aitchison, published in previous papers and updated to be the state-of-the-art at that time. All the basic principles are there: logratios, subcompositional coherence, the logistic-normal distribution, and many more ideas that are still inspiring present-day research. Many examples illustrate the methodology, most of the data invented by Aitchison himself, with a touch of his sense of humour in naming variables and data sets.

- Vera Pawlowsky-Glahn and Antonella Buccianti, editors (2011) *Compositional Data Analysis: Theory and Applications.* John Wiley, Chichester.

This multi-authored edited book, intended as a Festschrift for John Aitchison, presents a wide spectrum of the field of compositional data analysis, including a theoretical part, but — more useful for practitioners — many chapters on applications in research areas such as genomics, ecology, geochemistry and economics, as well as the use of an R package `robCompositions` for robust compositional data analysis

- K. Gerald van den Boogaart and Ramon Tolosona-Delgado (2013) *Analyzing Compositional Data with R.* Springer-Verlag, Berlin.

Probably the most suitable book for supporting this book and providing many more details, these authors use their R package `compositions` to perform many practical applications. In the process they explain all the basic principles as well as advanced topics in compositional data analysis. The book starts with an introduction to R, so is usefully self-contained in this respect.

- Vera Pawlowsky-Glahn, Juan José Egozcue and Raimon Tolosana-Delgado (2015) *Modeling and Analysis of Compositional Data*, Wiley.

The accent is on theoretical aspects, with only about 30% of the content involving real applications. Several theoretical issues not described in this book are treated: for example, distributional issues beyond the normal distribution and compositional processes. The authors' avoidance of amalgamations and additive logratios in practical applications marks a clear difference with this book, in which their usefulness in practice has been clearly demonstrated.

B.2 Articles

- Arnold van den Wollenberg (1977) Redundancy analysis. An alternative for canonical correlation analysis. *Psychometrika* 42:207–219.

Although the idea of redundancy analysis had already been described by C.R. Rao in a paper in 1964, called "principal component analysis with respect to instrumental variables", this paper by van den Wollenberg is regarded as the more accessible one, and the first to use the term "redundancy analysis", the more common term today.

- John Aitchison (1982) The statistical analysis of compositional data (with discussion). *Journal of the Royal Statistical Society, Series B (Methodology)* 44: 139–177.

Presented to the Royal Statistical Society, this masterful paper and its discussion make fascinating and obligatory reading for statisticians and applied researchers alike. It is interesting that, at this early stage, Aitchison admitted additive logratios and the operations of amalgamating and partitioning components (parts) of a composition, showing that he was aware of the obvious necessity of these operations in the practice of compositional data analysis.

- John Aitchison (1983) Principal component analysis of compositional data. *Biometrika* 79: 57–65.

This is Aitchison's first publication on what is called unweighted logratio analysis in the present book, namely the principal component analysis of centred logratios (cf. papers of Paul Lewi mentioned below). In this paper, his version of the logratio distance between samples is also defined.

- Wojtek Krzanowski (1987) Selection of variables to preserve multivariate structure. *Applied Statistics* 36: 22–33.

Krzanowski's idea was not to reduce dimensionality, but to eliminate variables that were contributing very little to the multivariate structure, an idea which has been exploited in the context of compositional data in Chap. 9. He used the Procrustes correlation as a measure of how well the structure was preserved, whereas in Chap. 9 the criterion is based on explained variance using redundancy analysis.

- John Aitchison and Michael Greenacre (2002) Biplots of compositional data. *Applied Statistics* 51: 375–392.

This is the paper, mentioned in the Preface, which was "rescued" and eventually published. It deals with biplots based on what is called unweighted logratio analysis in this book, where the compositional parts are weighted equally. This article precedes the introduction of differential part weights into logratio analysis — see the references to Paul Lewi and his spectral mapping below.

- Luc Wouters, Hinrich Göhlmann, Luc Bijnens, Stefan Kass, Geert Molenberghs and Paul Lewi (2003) Graphical exploration of gene expression data: a comparative study of three multivariate methods. *Biometrics* 59: 1131–1139.

One of the few papers that makes a direct comparison between principal component analysis, correspondence analysis and spectral mapping (alias weighted logratio analysis), in the context of a large gene expression data set.

- Juan José Egozcue, Vera Pawlowsky-Glahn, Gloria Mateu-Figueras and Carles Barcelo-Vidal (2003) Isometric logratio transformations for compositional data analysis. *Mathematical Geology* 35: 279–300.

The original article on isometric logratios.

- Paul Lewi (2005) Spectral mapping, a personal and historical account of an adventure in multivariate analysis. *Chemometrics and Intelligent Laboratory Systems* 77: 215–223.

This is another paper for obligatory reading by compositional data analysts, since it explains the history of spectral mapping (weighted logratio analysis in this book, or referred to simply as logratio analysis). Spectral mapping dates back to 1976 when Lewi published it in a now defunct medical journal, *Arzneimittel Forschung – Drug Research*. Paul Lewi, one of the founders of the discipline of chemometrics, independently (and probably before Aitchison) defined a multivariate dimension-reduction method that is based on a double-centred log-transformed table of positive data, but with the improvement that the analysis is weighted proportional to the row and column margins. The need for this weighting was inspired by Lewi's work in analysing biological activity spectra in drug development and by his knowledge of Benzécri's work in correspondence analysis, where row and column weights proportional to the respective margins are also used.

- Michael Greenacre and Paul Lewi (2009) Distributional equivalence and subcompositional coherence in the analysis of compositional data, contingency tables and ratio-scale measurements. *Journal of Classification* 26: 29–54.

This co-authored paper compares the respective approaches of John Aitchison and Paul Lewi on biplots of the relative (i.e. ratio-scale) values of positive data, showing that weighted logratio analysis (Lewi's spectral mapping) is not only subcompositionally coherent but also obeys the founding principle of correspondence analysis, namely distributional equivalence. The value is demonstrated of introducing weights to account for varying relative errors in the compositional parts.

- Michael Greenacre (2009) Power transformations in correspondence analysis. *Computational Statistics and Data Analysis* 53: 3107–3116.

This paper shows the link between logratio analysis (both unweighted and weighted) and correspondence analysis, a fact that has many practical consequences. The Box-Cox power transformation, which has the logarithmic transformation in the limit as the power parameter tends to zero, is the key to establishing the relationship between these two methods, which up to that time appeared unrelated. One of the consequences is that it provides an alternative approach to handling zero values in a compositional data matrix (see Chap. 8).

- Michael Greenacre (2011) Measuring subcompositional incoherence. *Mathematical Geosciences* 43: 681–693.

This paper provides a measure of how close a multivariate method is to being subcompositionally coherent, allowing a modest amount of incoherence to justify the use of the method as opposed to the ideal logratio approach that is perfectly coherent. Correspondence analysis is shown to be measurably close to being subcompositionally coherent (see Chap. 8).

- Michael Greenacre (2013) Contribution biplots. *Journal of Computational and Graphical Statistics*: 22, 107-122.

This paper explains the idea of the contribution biplot, used several times in this book. This is an innovation to make the coordinate lengths of the variables in a biplot be directly related to their contributions to the dimensions of the low-dimensional solution. Especially when there are many variables, this biplot helps the practitioner to focus on those variables that are important for the biplot interpretation.

- Javier Palarea-Albaladejo, Josep Antoni Martín-Fernández, Antonella Buccianti (2014) Compositional methods for estimating elemental concentrations below the limit of detection in practice using R. *Journal of Geochemical Exploration* 141: 71–77.

This article describes different ways of coping with the problem of zeros in compositional data, in the context of geochemistry.

- Angelina Kraft, Martin Graeve, D. Janssen, Michael Greenacre and Stig Falk-Petersen (2015) Arctic pelagic amphipods: lipid dynamics and life strategy. *Journal of Plankton Research*: 37, 790–807.

The case study using fatty acid compositional data in Chap. 10 of the present book is based on this paper, where more details about the study can be found.

- Michael Greenacre (2016) Data reporting and visualization in ecology. *Polar Biology*: 39, 2189-2205.

This article gives practical advice about reporting and graphing univariate and multivariate data. Several of the R functions described and illustrated in this paper are included in the `easyCODA` package.

- Michael Greenacre (2018) Variable selection in compositional data analysis. *Mathematical Geosciences*: in review.

This article is a more extensive treatment of the subject of variable selection (cf. Chap. 9).

- Martin Graeve and Michael Greenacre (2018) The selection and analysis of fatty acid ratios: A new approach for the univariate and multivariate analysis of fatty acid trophic markers in marine organisms. *Limnology and Oceanography: Methods*: in review.

This article is the application of the stepwise approach advocated in Chap. 9 of the present book, applied to two fatty acid data sets. The collaboration of a biochemist and a statistician leads to a selection of logratios suggested by the statistical criterion of explained variance, but the final choice is made according to substantive biochemical reasoning.

B.3 Web resources

- Supplementary material for this book is available at
  `https://www.crcpress.com/9781138316430`

The latest version of the `easyCODA` package can be downloaded here, which at any point in time might be updated compared to the version on CRAN. The data sets included in the package are given here, as well as a more detailed R script than described in Appendix C.

- R package `compositions`, by Gerald van den Boogaart and Raimon Tolosana-Delgado.

This very extensive package, which can be installed from CRAN, includes many additional functions and data sets — see the book by the same authors mentioned at the start of this Appendix.

- *Biplots in Practice*, by Michael Greenacre (2010), published by the BBVA Foundation, Bilbao. Freely downloadable at
  `www.multivariatestatistics.org`.

This online book treats biplots of all types: principal component analysis, correspondence analysis, logratio analysis, discriminant analysis, including centroid discriminant analysis which is used in Chap. 10 of the present book. Data sets and R code are available at the same website. In addition, the Spanish translation of *Correspondence Analysis in Practice. Second Edition* (Michael Greenacre, 2007; Spanish Edition, 2008) as well as the book *Multivariate Analysis of Ecological Data* (Michael Greenacre and Raul Primicerio, 2013), can be dowloaded for free from the same website.

- CODA Association debates at
  `www.coda-association.org/en/coda-info/coda-letters/`.

At present there is one debate, with many interchanges between the protractors and detractors of isometric logratios.

Appendix C
Computation of compositional data analysis

In this appendix the main features of the R code for performing compositional data analysis are summarized, including the use of the R package easyCODA, which accompanies this book. Commands in R are in red and always preceded with the prompt symbol >. Results are in blue. All lines of code are numbered on the right for possible referencing in the text. A selection of results are reported here, but a more extensive script is available as supplementary material.

C.1 Simple graphics for compositional data

In Chap. 1 the following small data set of ratios was plotted:

```
B6 1.576   6.333   4.019    0.884    0.561   0.140
B7 1.224   4.800   3.920    0.466    0.381   0.097
D4 2.184  16.462   7.538   25.680   11.760   1.560
D5 3.127  35.650  11.400   17.825    5.700   0.500
H5 0.504   0.418   0.830    3.540    7.020   8.460
H6 0.946   0.409   0.432    3.542    3.746   8.661
```

In the R code supplied online, these ratios as well as the original data set are included. It can be copied and then read in from the clipboard with the read.table command, which is slightly different depending on whether R is being run under Windows or MacOs. Having read in the data, the following code produces Fig. 1.1, using the function DOT included in the easyCODA package, which needs to be installed (in R: install.packages("easyCODA")) and loaded.

```
# Using Windows
> ratios <- read.table("clipboard", row.names=1)
# Using MacOs
> ratios <- read.table(pipe("pbpaste"), row.names=1)
# colorspace package for HCL colour palette
> require(colorspace)
# DOT function makes dot plot of ratios and logratios
> require(easyCODA)
> DOT(ratios, names=1:6, cols=rainbow_hcl(ncol(ratios)),
+     pch=rep(19, ncol(ratios)), ylim=c(0,40), ylab="Ratios")
> DOT(log(ratios), names=1:6, cols=rainbow_hcl(ncol(ratios)),
+     pch=rep(19, ncol(ratios)), ylim=c(-3,4), ylab="Logratios")
```

The Vegetables data set is also provided as part of the online R code, so suppose that this data set has been input to the R object veg. The function BAR plots the compositional bar chart of Fig. 2.2.

```
# Option order.column specifies which column orders the samples
# (if omitted, samples are plotted in orginal order)
> BAR(veg, order.column=2, ylab="Vegetables", cols=rainbow_hcl(ncol(veg)))
```

The ternary (or triangular) plot of Fig. 2.5 is obtained using the function ternaryplot in the package vcd:

```
# Ternary plot of Fig. 2.5
> require(vcd)
> par(mar=c(1,1,1,1))
> ternaryplot(veg, cex=0.6, col="blue", id=rownames(veg), main="",
+               dimnames_color="red", labels="outside")
```

C.2 Logratio transformations

Logratio transformations are computed by a series of functions in R package easyCODA, which should be installed:

- LR – computes all pairwise logratios
- ALR – computes additive logratios
- CLR – computes centred logratios
- ILR – computes one specific isometric logratio
- SLR – computes one specific amalgamation (i.e., summed) logratio
- PLR – computes set of pivot logratios

All of them have optional weights on the parts in the form of these possible options:

- weight = TRUE: weights equal to the means of the compositional parts, which is the default in all functions
- weight = FALSE: unweighted, i.e. weights all equal to 1 divided by the number of parts
- weight = specified set of positive weights (these will be closed to sum to 1).

Logratios as well as their respective weights are computed and returned by each the above functions, in $LR and $LR.wt respectively. The weights will depend on the weight option chosen.

In addition, there is a function to compute variances of any set of logratios and their total variance:

- LR.var — computes the variance of one or more logratios, usually the outputs from one of the above functions

LR.var can be used for any matrix, not just logratios. It produces the total variance, optionally weighted, of the columns of the input matrix, as well as the individual variances of the columns, if requested. When applied to a matrix of logratios, the function computes the total logratio variance.

The following is an example of the ALR function, showing partial output of its result. The denominator, used as the reference part, is by default the last part, but can be defined as any of the parts using the denom option, as illustrated here where denom = 2, the Carbohydrate column. Notice that the data frame veg is also available in the easyCODA package.

```
> require(easyCODA)
# Computing additive logratios
> data(veg)
> ALR(veg, denom=2)
$LR
                Protein/Carbohydrate Fat/Carbohydrate
Asparagus                 -0.5379911       -2.9272308
Beans(soya)                0.1586793       -0.4859608
Broccoli                   0.1062966       -1.7602821
   :                           :                 :
Spinach                   -0.2383746       -2.2305642

$LR.wt
Protein/Carbohydrate       Fat/Carbohydrate
          0.17410491             0.03702444

# to extract the ALRs to an object
> veg.ALR <- ALR(veg, denom=2)$LR
```

The output, which is the same for all these logratio functions, consists of two components: the matrix of computed logratios ($LR) and their associated weights ($LR.wt).

Now a series of examples follows, using mostly the Vegetables data set in R object veg, explaining the options of each logratio function and showing partial output. Notice again that the returned weights are according to the default option weight=TRUE — if equal weights are required, use the option weight=FALSE.

```
# Computing centred logratios
> CLR(veg)
$LR
                   Protein  Carbohydrate       Fat
Asparagus      -0.24817673    0.28981441 -2.6374164
Beans(soya)     0.14483568   -0.01384367 -0.4993515
Broccoli        0.17323521    0.06693863 -1.6933434
   :                :              :          :
Spinach        -0.06038357    0.17799100 -2.0525732

$LR.wt
     Protein  Carbohydrate         Fat
  0.24973580    0.69715641  0.05310779
```

```
# Computing an isometric logratio
# numer = numerator set, denom = denominator set
> ILR(veg, numer=1, denom=2:3)
$LR
         Asparagus       Beans(soya)        Broccoli         Carrots
        -0.14318395        0.08356201      0.09994693     -0.89661288
             Corn          Mushrooms           Onions            Peas
        -0.77331246       -0.58246827     -0.78674480     -0.31450146
Potatoes(boiled)           Spinach
        -0.90228421       -0.03483791

$LR.wt
Protein/Carbohydrate&Fat
              0.1873678
```

```
# Computing an amalgamation logratio
# numer = numerator set, denom = denominator set
# e.g. this SLR of first part relative to sum of parts 2 and 3
# (for both ILR and SLR, the two subsets can be simply indicated in order)
> SLR(veg, 1, 2:3)
$LR
          Asparagus        Beans(soya)         Broccoli           Carrots
         -0.25545319        -0.13890208       -0.02268706        -1.02031130
               Corn          Mushrooms           Onions              Peas
         -0.89102654        -0.69313881       -0.93033063        -0.43625982
   Potatoes(boiled)            Spinach
         -1.06669157        -0.14736751

$LR.wt
Protein/Carbohydrate&Fat
          0.1873678
```

```
# Computing a set of pivot logratios
# ordering = original order of parts by default
# but can be reordered as illustrated below
> PLR(veg, ordering=c(1,3,2))
$LR
                  Protein/Fat&Carbohydrate Fat/Carbohydrate
Asparagus                     -0.14318395       -0.6502705
Beans(soya)                    0.08356201       -0.1078533
Broccoli                       0.09994693       -0.3910384
   :                               :                :
Spinach                       -0.03483791       -0.4955093

$LR.wt
Protein/Fat&Carbohydrate         Fat/Carbohydrate
          0.18736783               0.03702444
```

```
# Computing all pairwise logratios
# order as in columns of upper triangle above diagonal of square matrix
# i.e. 1st/2nd, 1st/3rd, 2nd/3rd, 1st/4th, 2nd/4th, 3rd/4th, 1st/5th etc...
> LR(veg)
$LR
                  Protein/Carbohydrate Protein/Fat Carbohydrate/Fat
Asparagus                   -0.5379911   2.3892396        2.9272308
Beans(soya)                  0.1586793   0.6446402        0.4859608
Broccoli                     0.1062966   1.8665786        1.7602821
   :                             :           :                :
Spinach                     -0.2383746   1.9921896        2.2305642

$LR.wt
Protein/Carbohydrate         Protein/Fat         Carbohydrate/Fat
          0.17410491           0.01326292            0.03702444
```

```
# Plot of two amalgamation logratios and corresponding isometric logratios
# (cf. Figs 3.2 (a) and (b) that plots the ratios on a log-scale
# whereas here the logratio values are plotted directly)
# First simplify the names of beans and potatos
> rownames(veg)[c(2,9)] <- c("Beans", "Potatoes")
> amalgs <- cbind(SLR(veg, 3, 1:2)$LR, SLR(veg, 1,2)$LR)
> par(mar=c(4.2,4,3,1), mgp=c(2,0.7,0), font.lab=2)
> plot(amalgs, type="n", xlab="log(Fat/[Protein+Carbohydrate])",
+      ylab="log(Protein/Carbohydrate)", main="Amalgamations")
> text(amalgs, labels=rownames(veg))
> ilrs <- cbind(ILR(veg, 3, 1:2)$LR, ILR(veg, 1,2)$LR)
> plot(ilrs, type="n", xlab="sqrt(2/3)*log(Fat/[Protein+Carbohydrate]^0.5)",
+      ylab="sqrt(1/2)*log(Protein/Carbohydrate)", main="ILRs")
> text(ilrs, labels=rownames(veg))
```

Some logratio inverse functions are provided to take sets of logratios as imputs and bring them back to their original parts:

- `invALR` – inverts additive logratios
- `invCLR` – inverts centred logratios

For example, if ALRs are computed with respect to the last part, then the function `invALR` brings them back to the compositional part values. In the following, inversion is illustrated when the denominator is the second part, in which case this should be specified for the inversion, otherwise the parts will not come out in their original order. Note that the original compositional values of the `Vegetables` data set were given as percentages, but are returned as proportions after inversion.

```
# Inverting ALRs with respect to second part
> veg.alr <- ALR(veg, denom=2)$LR
> invALR(veg.alr, denom=2, part.names=colnames(veg))
               Protein Carbohydrate    Fat
Asparagus       0.3566       0.6107 0.0327
Beans(soya)     0.4205       0.3588 0.2207
Broccoli        0.4869       0.4378 0.0753
   :                :            :      :
Spinach         0.4157       0.5276 0.0567
```

```
# Computing total logratio variance and optionally individual variances
# e.g. total logratio variance of RomanCups data set, using 11 CLRs
# Data available in easyCODA package
> data(cups)
> LR.VAR(CLR(cups))
[1] 0.002339335

# Computing same total logratio variance, but using all 55 pairwise LRs

> LR.VAR(LR(cups))
[1] 0.002339335

# Computing total variance and individual variances

> LR.VAR(CLR(cups), vars=TRUE)
$LRtotvar
[1] 0.002339335

$LRvars
          Si            Al            Fe            Mg            Ca            Na
3.964377e-04  1.078177e-04  1.275227e-04  1.394376e-04  5.675271e-04  5.294847e-04
           K            Ti             P            Mn            Sb
1.045240e-04  2.045581e-05  1.456716e-05  1.507492e-05  3.164853e-04

# Percentage contributions to total logratio

> cups.var <- LR.VAR(CLR(cups), vars=TRUE)
> round(100 * cups.var$LRvars / cups.var$LRtotvar, 2)
   Si    Al    Fe    Mg    Ca    Na     K    Ti     P    Mn    Sb
16.95  4.61  5.45  5.96 24.26 22.63  4.47  0.87  0.62  0.64 13.53

# Percentages for unweighted data, showing the dominance of the rarest elements,
# particularly Mn which only has values (in percent) 0.01, 0.02 and 0.03

> cups.var <- LR.VAR(CLR(cups, weight=FALSE), vars=TRUE)
> round(100 * cups.var$LRvars / cups.var$LRtotvar, 2)
   Si    Al    Fe    Mg    Ca    Na     K    Ti     P    Mn    Sb
 7.69  2.48  5.92  6.42  1.93  3.63  4.17  4.93  5.36 29.16 28.32
```

C.3 Compositional data modelling

Various standard functions in R, e.g. `lm` (linear modelling), `glm` (generalized linear modelling), and `rpart` (recursive partitioning, or classification and regression trees) can be used to build models using compositional data. For example, using the FishMorphology data set, stored in R object `fish`, the following code determines the best predictor of the fish's sex, using logistic regression (see Sect. 5.2), and evaluates the classification accuracy.

```
# Predicting sex of fish in data set FishMorphology
# fish data frame in easyCODA has sex, habitat and mass in first three columns
> data(fish)
> sex      <- fish[,1]
> habitat <- fish[,2]
> mass     <- fish[,3]

# remaining columns contain fish morphometric data, in object 'fishm' (75x26)
> fishm <- as.matrix(fish[,4:29])

# convert fishm to a compositional data matrix
> fishm <- fishm/apply(fishm, 1, sum)

# compute all logratios
> fish.LR <- LR(fishm)$LR

# look for best logratio predictor of sex
# save deviance measures of lack of fit in 'dev'
# numLR = total number of logratios
> numLR <- ncol(fishm) * (ncol(fishm)-1) / 2
> dev     <- rep(0, numLR)
> for(j in 1:numLR) dev[j] <- summary(glm(factor(sex) ~ fish.LR[,j],
+                             family="binomial"))$aic
> j.min <- which(dev == min(dev))
> colnames(fish.LR)[j.min]
[1] "Faw/Fdl"
> fish.glm <- glm(factor(sex) ~ fish.LR[,j.min], family="binomial")
> summary(fish.glm)
:
:
Coefficients:
                Estimate Std. Error z value Pr(>|z|)
(Intercept)       -3.687      1.083  -3.403 0.000666 ***
fish.LR[, j.min] -12.417      3.522  -3.525 0.000423 ***
---
:
:
    Null deviance: 103.959  on 74  degrees of freedom
Residual deviance:  88.292  on 73  degrees of freedom
AIC: 92.292
# (no other logratios add significantly to this model)

# Predictions of log-odds: positive predictions (i.e. p>0.5) are for males
> fish.pred <- predict(fish.glm)
> table(fish.pred>0, factor(sex))
          1  2
  FALSE 26 10
  TRUE  11 28
```

The alternative approach of classification trees (see Fig. 5.3) uses the R package and function `rpart`. It is computed and plotted as follows, along with its classifaction accuracy, showing it does better predicting the female fish and almost as good predicting the males.

```
# Classification tree, predicting sex from morphometric logratios
> require(rpart)
> fish.tree <- rpart(factor(sex) ~ fish.LR)
> plot(fish.tree, margin=0.1)
> text(fish.tree, use.n=T)
> fish.pred.tree <- predict(fish.tree)
> table(fish.pred.tree[,1]<0.5, factor(sex))
        1  2
  FALSE 36 11
  TRUE   1 27
```

When the compositional data set forms a set of multivariate responses, then function `rda` for redundancy analysis (RDA) in the `vegan` package can be used (later the function RDA, which is part of the `easyCODA` package, will be used to map RDA results). Here is some code that computes the results in Sect. 5.4, where the logratio variance explained by the logarithm of fish mass is computed and tested.

```
# fishm contains morphometric measurements (see previous page)
# Scale the CLRs with square root of their weights
# to ensure weighted logratio variance is explained
> fish.CLR <- CLR(fishm)
> fish.CLRw <- fish.CLR$LR %*% diag(sqrt(fish.CLR$LR.wt))

# logmass contains the logarithm of fish masses
> logmass <- log(mass)
> require(vegan)
> fish.rda <- rda(fish.CLRw ~ logmass)
> set.seed(123)
> anova(fish.rda, by="terms")
:
Number of permutations: 999
Model: rda(formula = fish.CLRw ~ logmass)
          Df    Variance      F Pr(>F)
logmass    1 0.00007157  2.717  0.002 **
Residual  73 0.00192286
```

C.4 Compositional data analytics

The following functions are provided in `easyCODA` for performing various multivariate analyses of compositional data:

- LRA – logratio analysis, with plot function PLOT.LRA
- PCA – principal component analysis, with plot function PLOT.PCA
- CA – correspondence analysis, with plot function PLOT.CA
- RDA – redundancy analysis, with plot function PLOT.RDA
- STEP – stepwise selection of logratios (or any independent variables)

Note that it is sufficient to use the function PLOT for the plotting of results of LRA, PCA, CA and RDA — PLOT will automatically recognize the respective object. Function RDA has some differences compared to `rda` in the `vegan` package, mainly that it is written in the same style as the other multivariate functions in the `easyCODA` package.

These functions all depend on the `ca` package, which needs to be installed in R: `install.packages("ca"))` and then loaded during each session. For this reason the generic `plot` and `summary` functions from the `ca` package can also be used.

Logratio analysis (LRA)

The function `LRA` performs logratio analysis (LRA), and includes the definition of amalgamation logratios specified in a list. If an amalgamation logratio is defined, all its component parts are automatically declared *supplementary* points, equivalently known as *passive* points, as opposed to the *active* points that determine the principal axes of the solution. There is also an option to make the aggregations supplementary, while maintaining their component parts active. There are two options for plotting the results: first the generic `plot` function from the `ca` package (remember that LRA, PCA and CA all generate `ca` objects), and second the `PLOT.LRA` function that is part of the `easyCODA` package. Here are two examples, the first one reproduces the two plots in Fig. 6.1 on the `RomanCups` data (data object `cups` in `easyCODA`), using `plot` from the `ca` package, i.e. `plot.ca`. The latter function has an option for showing symbols proportional to row and column weights, called "masses" in correspondence analysis, used in the weighted LRA plot in Fig. 6.1.

```
# LRA, weighted and unweighted, of the RomanCups data set
# plotting by ca package function (default is symmetric plot)
# the mass option shows column symbols related to their weight
> require(ca)
> data(cups)
> cups.wLRA <- LRA(cups)
> cups.uLRA <- LRA(cups, weight = FALSE)
> par(mar=c(3.5,3.3,4,2), font.lab=2, mgp=c(2,0.5,0), cex.axis=0.8)
> plot(cups.uLRA, labels=c(1,2), main="Unweighted LRA")
> plot(cups.wLRA, labels=c(1,2), mass=c(FALSE,TRUE), main="Weighted LRA")
```

The second example uses `PLOT.LRA` from `easyCODA`,which has a rescaling option, applied to the `TimeBudget` data.

```
# The TimeBudget data set is in the R object 'time', available in easyCODA
# Usually, weighted LRA will be used
> data(time)
> time.LRA <- LRA(time)
> par(mar=c(3.5,3.3,4,2), font.lab=2, mgp=c(2,0.5,0), cex.axis=0.8)
> PLOT.LRA(time.LRA, map="contribution", main="No rescaling")
> PLOT.LRA(time.LRA, map="contribution", rescale=0.5,
+          main="Rescaling columns by 0.5")
```

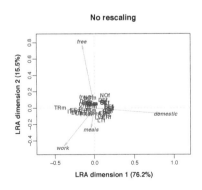

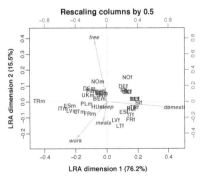

In symmetric maps the row and column points have equal dispersions along the axes, but the asymmetric and contribution biplots often have one set of points much less dispersed than the other, which affects legibility, as seen in the first plot. PLOT.LRA allows for different scales for the rows and columns, as shown in the second plot, where the scaling factor (option `rescale`) is applied to the column coordinates. In this case there are two sets of tick marks for the respective points, the tick marks at the top and on the right referring to the column points.

Since the object created by LRA is a ca object, the generic `summary` function in the ca package can be used to obtain all the numerical results of the LRA. Here is a partial listing, omitting some of the results for the samples, for the `time.LRA` object computed on the previous page:

```
> summary(time.LRA)

Principal inertias (eigenvalues):

 dim     value      %    cum%    scree plot
  1     0.025449  76.2   76.2    ********************
  2     0.005168  15.5   91.7    ****
  3     0.001871   5.6   97.3    *
  4     0.000677   2.0   99.3    *
  5     0.000229   0.7  100.0
        --------  -----
 Total: 0.033394 100.0

Rows:
        name   mass  qlt   inr    k=1  cor  ctr      k=2 cor ctr
  1  |   TRm |   31   964  184 |  -436  964 233 |     4   0   0 |
  2  |   BEm |   31   409    6 |   -48  357   3 |    18  52   2 |
  3  |   DEm |   31   974   19 |  -113  634  16 |    83 341  41 |
  4  |   FRm |   31   791   21 |  -109  532  15 |   -76 259  35 |
  :  |    :  |    :     :    : |    :    :    : |     :   :   : |
 32  |   SIf |   31   964   40 |   204  963  51 |    -7   1   0 |

Columns:
        name   mass  qlt   inr    k=1  cor  ctr      k=2 cor ctr
  1 | work |   213   909  180 |  -141  703 166 |   -76 206 239 |
  2 | dmst |   115   996  608 |   419  995 794 |   -16   1   5 |
  3 | trvl |    57   364   34 |   -74  275  12 |    42  88  19 |
  4 | slep |   341   142    6 |    -2   10   0 |    -9 132   5 |
  5 | mels |    97   262   48 |   -24   36   2 |   -61 226  70 |
  6 | free |   177   980  125 |   -61  160  26 |   139 820 661 |
```

After the summary of the variances on each dimension (the eigenvalues or squared singular values, called "principal inertias" in correspondence analysis), there is a numerical listing for the rows and the columns. All values are in thousandths (per-mills) or multiplied by a 1000. The column `mass` gives the weights of the row and column points, the column `qlt` gives the quality (R^2) of each point's explanation by the two listed dimensions (principal axes), and the column `inr` gives the contributions of each point to the total variance. The columns `k=1` and `k=2` give the principal coordinates on the first two dimensions, the columns `cor` give squared correlations with the respective axes, and the columns `ctr` give the contributions to the axes. The column `qlt` is the sum of the `cor` columns.

A LRA can be performed with amalgamations specified as a list, usually for a substantive research reason. For example, suppose that sleep, meals and free time are required to be aggregated, called 'nowork', to be contrasted with the other three activities. The following LRA could then be performed (automatically, the individual parts sleep, meals and free will be made supplementary columns):

```
# LRA with amalgamation specified
> LRA(time, amalg(nowork=c(4:6)))
```

Principal component analysis (PCA)

The PCA function for principal component analysis (PCA) includes weighting of both rows and columns, and also allows supplementary rows and columns, which is not typically available in R. Here are some examples, using the plot function PLOT.PCA:

```
# compute logratios of Vegetables data and perform unweighted PCA
> veg.LR <- LR(veg)
> veg.PCA <- PCA(veg.LR$LR, weight = FALSE)
> PLOT.PCA(veg.PCA, map="asymmetric")
```

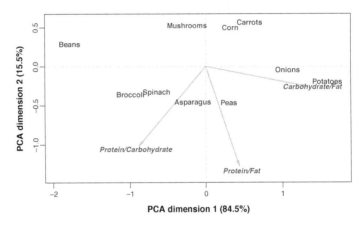

Compare the plot on the next page with the left hand plot in Fig. 3.2(d) — both axes have been reversed compared to that plot and here the logratios are also represented in the biplot.

The next example, a weighted PCA of the RomanCups shows how to change the orientation of the axes (see plot on next page).

```
# perform weighted PCA on the ALRs, with denominator silicon
> cups.ALR <- ALR(cups, denom = 1)
# ALR column weights automatically picked up by the PCA
> cups.PCA <- PCA(cups.ALR)
# reverse second axis to agree with orientation of results in Chap. 9
> PLOT.PCA(cups.PCA, map="contribution", rescale=0.2, axes.inv=c(1,-1))
```

Redundancy analysis (RDA)

The RDA function for redundancy analysis (RDA) performs a PCA with constraints on the solution. When applied to the CLRs, it performs a constrained form of LRA. It can also be applied to a subset of the logratios, include weighting of the columns and /or rows. The function also allows supplementary rows and columns, like the

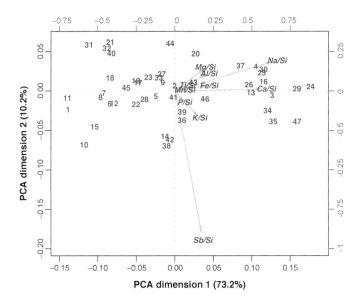

PCA and CA functions. Constraints can be applied using both continuous and categorical variables, the latter being converted to dummy variables using the function DUMMY. Here is how the contribution biplot in Fig. 6.4 was obtained using the FishMorphology data set.

```
# Perform RDA on the weighted CLRs of FishMorphology data set
# constraining variables are in matrix object vars
# assume fishm, mass, habitat and sex already available objects
# (see code at start of Section C.3)
> logmass <- log(mass)
> sexhab <- 2*(sex-1) + habitat
> sexhab.names <- c("FL","FP","ML","MP")
> rownames(fishm) <- sexhab.names[sexhab]

# create dummy variables for sexhab and create vars
> sexhab.Z <- DUMMY(sexhab, catnames=sexhab.names)
> vars <- cbind(logmass, sexhab.Z)

# perform RDA
> fish.RDA <- RDA(CLR(fish), cov=vars)

# plot RDA - note that option indcat indicates which columns
# in the constraining variables object 'vars' are dummies
> require(colorspace)
> colcat <- rainbow_hcl(4, c=80)
> par(mar=c(3.5,3.3,2.5,2.5), font.lab=2, mgp=c(2,0.5,0), cex.axis=0.8)
> PLOT.RDA(fish.RDA, map="contribution", rescale=0.02, axes.inv=c(1,-1),
+          indcat=2:5, colrows=colcat[sex_hab], colcats=colcat,
+          pchrows=rep(19, nrow(fishm)), cexs=c(1,1,1), fonts=c(2,4,2))
> legend("topright", legend=c("female litoral (FL)", "male litoral (ML)",
+        "female pelagic (FP)", "male pelagic (MP)"), pch=19,
+        col=colcat, text.col=colcat, text.font=2)
```

Fig. 6.5 shows a different representation for the samples (fish), as the weighted averages of the variable points that are now in standard coordinates (asymmetric biplot) This alternative representation also exists in the `vegan` package. In addition, confidence ellipses for the means of the four groups of points are added using the function `CIplot_biv` available online (see Greenacre 2006) and the link in the bibliography, web resources), but also included in the `easyCODA` package.

```
# First compute row points as supplementary weighted averages
# of column standard coordinates, then substitute them in RDA object
> fish.spc <- fishm %*% fish.RDA$colcoord
> fish.RDA$rowpcoord <- fish.spc
> par(mar=c(3.5,3.3,2.5,2.5), font.lab=2, mgp=c(2,0.5,0), cex.axis=0.8)
> PLOT.RDA(fish.RDA, map="asymmetric", rescale=0.02, axes.inv=c(1,-1),
+           indcat=2:5, colrows=colcat[sexhab], colcats=colcat,
+           pchrows=rep(19, nrow(fishm)), cexs=c(1,1,1), fonts=c(2,4,2))

# add confidence ellipses (needs package ellipse)
> require(ellipse)
> CIplot_biv(fish.spc[,1], -fish.spc[,2],  alpha=0.99, group=sexhab,
+            groupcols=colcat, shade=TRUE, add=TRUE, shownames=FALSE)
```

Procrustes analysis

Procrustes analysis can be performed by the functions `procrustes` and `protest` in the `vegan` package. Function `procrustes` rotates a configuration to maximum similarity with another configuration, while function `protest` computes the Procrustes correlation and performs a permutation test for non-randomness between two configurations.

As an example, the weighted and unweighted logratio analyses of the data set `RomanCups` on page 104 are compared with each other as well as with the correspondence analysis. The functions LRA and CA both output the row principal coordinates as components `$rowpcoord`. Procrustes correlations are computed between the row principal coordinates from each analysis — these coordinates yield weighted and unweighted logratio distances and chi-square distances respectively, in the full space of the solution. To just get the Procrustes correlation, the permutation test in `protest` can be suppressed by setting `permutations=0`.

```
# three sets of principal coordinates: unweighted LRA, weighted LRA and CA
> data(cups)
> cups.ulra.rpc <- LRA(cups, weight=FALSE)$rowpcoord
> cups.wlra.rpc <- LRA(cups)$rowpcoord
> cups.ca.rpc   <- CA(cups)$rowpcoord

# Procrustes correlations in full 10-D space
> require(vegan)
# ... between unweighted and weighted LRA
> protest(cups.ulra.rpc, cups.wlra.rpc, permutations=0)$t0
[1] 0.7377072
# ... between weighted LRA and CA
> protest(cups.wlra.rpc, cups.ca.rpc, permutations=0)$t0
[1] 0.9962435

# Procrustes correlations in reduced 2-D space
# ... between unweighted and weighted LRA
> protest(cups.ulra.rpc[,1:2], cups.wlra.rpc[,1:2], permutations=0)$t0
[1] 0.6979616
# ... between weighted LRA and CA
> protest(cups.wlra.rpc[,1:2], cups.ca.rpc[,1:2], permutations=0)$t0
[1] 0.9938629
```

Both in the full space and the reduced two-dimensional space, the agreement between weighted LRA and CA is very high, but not so for the weighted and unweighted LRAs, as was seen in the two-dimensional solutions of Fig. 6.1.

Stepwise selection of logratios

Variable selection is performed by the function STEP. This function aims to explain
the variance of a target matrix, in our case almost always the original compositional
data set. It has available to it another data matrix on which logratios will be constructed, often the same matrix, but could also include amalgamations.

The results for the RomanCups application in Chap. 9 were obtained by setting
a random number seed and then giving a simple command, as follows.

```
> set.seed(2872)
> cups.step <- STEP(cups, random=TRUE)
> cups.step
$R2max
 [1] 0.6153246 0.7411887 0.8641640 0.9358403 0.9661994 0.9844508
 [7] 0.9919763 0.9950725 0.9978274 1.0000000

$ratios
       row col
 [1,]   1   5
 [2,]   1  11
 [3,]   6  11
 [4,]   1   3
 [5,]   7  11
 [6,]   1   4
 [7,]   2   6
 [8,]   5   8
 [9,]   1  10
[10,]   9  11

$names
 [1] "Si/Ca" "Si/Sb" "Na/Sb" "Si/Fe" "K/Sb"  "Si/Mg"
 [7] "Al/Na" "Ca/Ti" "Si/Mn" "P/Sb"

$logratios
        Si/Ca    Si/Sb    Na/Sb    Si/Fe        K/Sb    Si/Mg     Al/Na  ...
1    2.710713 5.341802 3.812816 5.667225 0.20067070 5.075174 -2.181400  ...
2    2.508950 5.390869 4.010084 5.555172 0.28768207 5.058735 -2.313635  ...
3    2.401765 5.068046 3.791395 5.163356 0.20479441 4.809185 -2.238047  ...
 :     :        :        :        :        :          :         :         :
45   2.651119 5.499897 4.083171 5.723040 0.60613580 5.290176 -2.221678  ...
46   2.526775 5.214099 3.848510 5.139991 0.46357274 4.907369 -2.203804  ...
47   2.314196 4.800564 3.532533 5.154736 0.08408312 4.931592 -2.257465  ...

$pro.cor
 [1] 0.7844263 0.8572142 0.8620142 0.9003110 0.9026684 0.9175735
 [7] 0.9289906 0.9300123 0.9322281 0.9324183
```

The results of STEP are the cumulative explained variances (in $R2max, reported
in Table 9.2), the indices of the 10 logratios (in $ratios), the ratio names (in
$names), the complete logratio matrix, 47×10 in this example (in logratios)
and the Procrustes correlations in the stepwise process (in $pro.cor).

There can often be logratios that are tied for being selected, in terms of additional
variance explained, because of the interconnectedness of the logratios. For example,
in the first step above (see Table 9.2), the ratio *Si/Ca* was chosen, explaining 61.5%
of the logratio variance. In the second step the element *Sb* can enter either in a
ratio with *Si* or with *Ca*, and both explain an additional 12.6% of the variance (only

one can enter, otherwise a cycle is created — see Fig. 5.1(b) — and the solution is necessarily acyclic). The option `random=TRUE` indicates that such ties should be broken randomly, and in this case the ratio *Si/Sb* entered.

To obtain the contained variances in Table 9.2, express the variance of each logratio as a percentage of the total variance.

An enhancement of the algorithm is to break the ties by selecting the logratio that gives the highest improvement in the Procrustes correlation. A different sequence of logratios is then obtained, but the new elements entering at each step are the same. In this particular example, the optimal sequence turns out to be exactly the set of ALRs with respect to *Si*, reported as the first row of Table 9.1.

```
> STEP(cups)$names
 [1] "Si/Ca" "Si/Sb" "Si/Na" "Si/Fe" "Si/K"  "Si/Mg"
 [7] "Si/Al" "Si/Ti" "Si/Mn" "Si/P"
```

When amalgamations are added to the set of simple logratios, then the sequence of accumulated variances explained is as good as or better than before, except for the last step, the tenth step in this example. The explained variance does not reach exactly 100% in 10 steps, but in practice it is extremely close to explaining all the variance. Here all two-part amalgamations are added to the data set to explain the same total variance of the `RomanCups` data. The script reproduces the results of Table 9.4.

```
# compute amalgamations
> amalgs <- matrix(0, nrow(cups), ncol=0.5*ncol(cups)*(ncol(cups)-1))
> colnames(amalgs) <- 1:ncol(amalgs)
> cupnames <- colnames[cups]
> k <- 1
> for(jj in 2:ncol(cups)) {
>   for(j in 1:(jj-1)) {
+      amalgs[,k] <- cups[,j]+cups[,jj]
+      colnames(amalgs)[k] <- paste(cupnames[j], cupnames[jj], sep="+")
+      k <- k+1
+   }
> }

# combine amalgamations with original parts
> cups_amalgs <- cbind(cups, amalgs)

# perform stepwise selection and show some results
> cups.step2 <- STEP(cups_amalgs, cups)
> cups.step2$R2max
 [1] 0.6454359 0.7755263 0.8828579 0.9428255 0.9732152 0.9859494
 [7] 0.9914720 0.9950723 0.9976510 0.9994871
> cups.step2$names
 [1] "Si/Ca+Na"    "Na/P+Sb"     "Al+Mg/Na+Sb" "Na/Ca+P"     "K+P/Mg+Sb"
 [6] "Al+Fe/Mg+Ti" "Al+P/Fe+Mn"  "Fe+K/Ti+P"   "Mn/Fe+Ti"    "Ti/P"
```

Appendix D
Glossary of terms

In this appendix an alphabetical list of the most common terms used in this book is given, along with a short definition of each. Words in italic font that are used in the definitions refer to terms which are themselves contained in the glossary.

- *acyclic connected graph* — also called a tree, this is a *graph* in which two *vertices* are connected by exactly one path.

- *additive logratios (ALRs)* — a set of *logratios* where one compositional *part* serves as the denominator of each ratio, with the other *parts* being the numerators.

- *amalgamation* — a sum of the compositional values of selected *parts*, usually aggregated with a specific substantive intention.

- *amalgamation, or summed, logratio (SLR)* — a *logratio* where the numerator and/or the denominator is a sum of *parts*.

- *asymmetric biplot* — a *biplot* where one set of points is in *principal coordinates* and the other set in *standard coordinates*.

- *balance* — alternative name for an *isometric logratio* or *amalgamation logratio*, especially when several of these logratios are identified with a recursive partitioning of the set of *parts*.

- *barycentre* — synonym for *weighted average*.

- *biplot* — a joint display of rows and columns of a table, which approximates data values by scalar products.

- *centred logratio (CLR)* — a *logratio* with a *part* in the numerator and the *geometric mean* of all the *parts* in the denominator.

- *close* — to express a set of nonnegative values relative to their total; often the term "normalize" is used as a synonym.

- *composition* — a set of nonnegative real-valued numbers that add up to fixed total, usually 1 or 100%.

- *compositional data* — a set of *compositions* observed on several sampling units.

- *compositional data analysis* — the special area of Statistics devoted to the analysis of *compositional data*.

• *confidence ellipse* — the generalization of a confidence interval to a bivariate mean.

• *contingency ratio* — the ratio of a value in a two-way table to its expected value, where the expected value is usually based on the product of the marginal values.

• *contribution biplot* — a *biplot* where the variable arrows have lengths related to their contributions to the solution.

• *correspondence analysis (CA)* — a method of displaying the rows and columns of a table as points in a spatial map, with a specific geometric interpretation of the positions of the points as a means of interpreting the similarities and differences between rows, the similarities and differences between columns and the association between rows and columns.

• *detection limit* — the lowest quantity of a substance that can be distinguished from its absence.

• *edge* — (of a *graph*) the displayed connection between two *vertices*.

• *eigenvalue* — in several multivariate dimension-reducing methods, the maximized amount of variance accounted for by a dimension; equal to the squared *singular value*.

• *Euclidean distance* — (or Pythagorean distance) the distance equal to the square root of the sum of squared differences between corresponding coordinates of multivariate vectors; the squared differences are often weighted.

• *geometric mean* — the p-th root of the product of p parts.

• *graph* — a mathematical structure that models pairwise relationships between objects, called *vertices*, connecting them by *edges*.

• *inertia* — synonym for variance, especially when weights are incorporated.

• *isometric logratio* — a *logratio* where the numerator and/or the denominator is a *geometric mean* of *parts*; abbreviated as ILR. Associated with an ILR is a scaling constant which depends on the numbers of *parts* or the *weights* assigned to the parts, as the case may be.

• *logratio* — the logarithm of the ratio of two positive numbers, usually two *parts* of a *composition*.

• *logratio analysis (LRA)* — an approach related to *correspondence analysis*, but applied to the logarithms of a table with strictly positive values; the weighted form of LRA, where rows and columns are weighted, as in CA, is usually preferred over the unweighted form, and is taken as the default one in this book.

• *logratio distance* — the distance between samples or between rows equal to the Euclidean distance, usually weighted, computed on all pairwise *logratios*, or equivalently on the set of *centred logratios*.

• *part* — one of the components of a *composiiton*.

• *permutation test* — a distribution-free strategy of statistical inference achieved through the generation of data permutations, either all possible ones or a large random sample, assuming a null hypothesis, leading to the null distribution of a test statistic and thus an estimate of the p-value for the observed value of the statistic.

• *pivot logratios (ALRs)* — a set of *isometric logratios* where each ratio is of a part with respect to the *geometric mean* of the remaining parts in on ordered list.

• *principal axis* — a direction of spread of points in multidimensional space that optimizes the *variance* displayed; can be thought of equivalently as an axis which best fits the points in a weighted least-squares sense.

• *principal component analysis (PCA)* — a dimension-reducing method of multivariate analysis, where the dimensions explaining a maximum amount of variance are identified.

• *principal coordinates* — coordinates of a set of points projected onto a *principal axis*, such that their weighted sum of squares along an axis equals the *principal variance* on that axis.

• *Procrustes analysis* — a multivariate method that translates, rescales and rotates one multidimensional configuration of a set of objects to match as closely as possible another one of the same objects.

• *Procrustes correlation* — the measurement of agreement between two multidimensional configurations of the same objects, using *Procrustes analysis*.

• *redundancy analysis (RDA)* — a *principal component analysis* where the solution dimensions are constrained to be linearly dependent on a set of explanatory variables.

• *reference range* — a summary of the dispersion of a sample of values, based on lower and upper estimated percentiles, e.g. 2.5% and 97.5% for a 95% reference range.

• *simplex* — a triangle in two dimensions, a tetrahedron in three dimensions and generalizations of these geometric figures in higher dimensions; J-part *compositions* (called profiles in CA) lie inside a simplex defined by J vertices in a $(J-1)$-dimensional space.

• *singular value decomposition (SVD)* — a matrix decomposition of a rectangular matrix; the squares of the singular values are *eigenvalues* of particular square matrices, and the left and right singular vectors are also eigenvectors.

• *spectral mapping* — a synonym for weighted *logratio analysis*.

• *standard coordinates* — coordinates of a set of points such that their weighted sum of squares along an axis equals 1.

• *subcomposition* — a reduced set of parts of a *composition*, where the parts are re-closed to sum to 1.

• *subcompositional coherence* — a desirable property of *compositional data analysis* whereby relationships between the parts of a *subcomposition* remain the same as in an extended *composition*.

• *subcompositional incoherence* — a measure of how far a method of *compositional data analysis* is from exact *subcompositional coherence*.

• *symmetric map* — a joint display of the rows and columns where the two clouds of points have the same normalization in *principal coordinates* ; strictly speaking, the symmetric map is not a *biplot*.

• *vertex* — (in a *graph*) one of the objects being connected in a *graph*.

• *vertex* — (in a *simplex*) one of the "corners" of a *simplex* that is connected to other vertices defining the outer limits of the simplex space.

• *Ward clustering* — a specific hierarchical clustering algorithm that minimizes the within-cluster variance at each clustering step, equivalent to maximizing the between-cluster variance.

• *weight* — a positive value assigned to each row and each column of the *compositional data* table. Each set of weights usually sums to 1 (i.e. they themselves form a *composition*). In *correspondence analysis* the term "mass" is used for weight.

• *weighted average* — an average where differential *weights* are applied to each value being averaged.

Appendix E
Epilogue

In this final part of *Compositional Data Analysis in Practice*, I can finally use the first person singular again and express some personal comments and opinions about this subject, after working more than 45 years in the allied area of correspondence analysis and 18 years since my fortuitous meeting with John Aitchison.

Compositional data analysis is simple

As I mentioned in the Preface, in 1997 John Aitchison gave a talk with the title "The one-hour course in compositional data analysis, or compositional data analysis is simple", and I think he was perfectly correct. It is simple — it involves some simple transformations of the compositional parts, or components, of the compositions, after which it is statistical business as usual. The idea that a ratio of two compositional parts is invariant to the chosen set of parts (i.e. subcompositional coherence) is simple. The logarithm of a ratio, which puts ratios on an interval scale, is simple and common practice, exemplified by: $\log(a/b) = \log(a) - \log(b)$. The logratio transformation is even a familiar one to most statisticians and econometricians, thanks to the logit, or log-odds, transformation in logistic regression, or the definition of a financial return as the logarithm of a price ratio, for example. And then, when parts are to be combined, for example to obtain the proportion of saturated fatty acids in a sample where many different fatty acids have been measured compositionally, the corresponding part values are simply added up — a straightforward amalgamation.

This introductory book on compositional data analysis is quite short and could probably be even shorter. Once one understands these simple basics, then the rest is just carrying on with the usual statistical paraphernalia. There is no need to repeat what already exists in scores of textbooks on multivariate analysis. You want to do a regression where compositional data are regarded as predictors? Don't use their original values, use logratios of the compositional parts. You want to do a cluster analysis of samples based on their respective compositions, or a canonical correlation analysis between two sets of compositions on the same samples? Again, don't use the original data, transform to logratios first and then do it. Aitchison's one-hour course is probably all you need, plus a regular education and understanding of modern statistics or data science.

There are some special features, of course, that are unique to compositional data, such as the problem with data zeros and the constant sum constraint. This constraint filters through to the logratios in the form of an interesting network of exact linear relationships: for a composition consisting of $J = 10$ parts, for example, there are only 9 logratios that are needed to generate the values of all the 55 logratios of pairs of parts ($55 = 10 \times 9/2$). This fact has further repercussions when constructing covariance matrices between logratios and when looking for important logratios in modelling.

So why make it complicated?

Compositional data exist in a simplex, but the logratio transformation takes the data out of this restricted space into unbounded real space, where differences between values are measured on a regular interval scale and where familiar statistical analysis, both univariate and multivariate, can be performed. A similar situation exists for observed positive data in many research fields, for example counts in sociology and stock prices in finance, where values are compared percentagewise. For these types of data, a log-transformation converts them to interval-scale measurements, where regular methods such as regression and analysis of variance can be applied, since these methods depend on interval-scale differences.

In log-linear modelling and Poisson regression of count data, for example, there has never been any attempt to develop, nor need to develop, a ratio-scale algebra in the space of the original counts. Yet in compositional data analysis, such an algebra has been developed for the original compositional data in the simplex, without any clear need for this from a practical viewpoint. In my opinion, this algebra is more of a hindrance to the practice of compositional data analysis than a benefit.

For example, consider the simple statement made above, which in general is that, for a J-part composition, only $J - 1$ of them are needed to generate any of the $J \times (J-1)/2$ logratios. Any of the sets of $J - 1$ additive logratios (ALRs) will do the job, and this result is trivial. In the book by Pawlowsky-Glahn et al. (2015), much store is placed on simplex algebra and redefining operations such as addition and scalar multiplication in the logratio world by "perturbation" and "powering" in the simplex one. It is not at all clear what the benefit is of all this "simplicial mathematics" in parallel with the regular algebra and geometry in the logratio-transformed space. From a practitioner's point of view, I would say there is no advantage at all, it just redefines what is basically simple mathematics as a complex system with new terminology, new notation and new expressions.

One reason for this "mathematicizing" of the subject seems to be to wrap it in a framework called the "Aitchison geometry", a framework from which it is difficult to escape due to its algebraic rigour. Called the logratio space in this book, it is simply understood as the Euclidean geometry of logratio-transformed compositions, and includes the definition of the "Aitchison distance", called the logratio distance in this book, both between samples (usually the rows of a compositional data matrix) and between parts (usually the columns). The logratio distance space can certainly be held up as one type of ideal, since it contains the totality of logratio variance. But like any multivariate space in general, it includes measurement error as well as a high portion of variance that can be considered to be random fluctuations in the data, i.e. what is "noise" rather than "signal" (see Chapter 9, Section 9.5).

Hence, the approach taken in this book is to try to identify those logratios that are explaining a large part of the logratio variance, admitting that not all of this variance is relevant to the problem at hand. Simplicity, parsimony and ease of substantive interpretation are the hallmarks of this approach, rather than trying to stay rigorously within the "Aitchison geometry" using unnecessarily complicated concepts.

The ILR transformation

Speaking of complicated concepts, I have taken a stand against using the isometric logratio (ILR) as a transformation of compositional data, because of its problematic interpretation. I realize that this is a controversial viewpoint, since much investment has been made in its diffusion and usage. But even John Aitchison himself expressed this view as early as 2003 in a talk (cited by Egozcue and Pawlowsky-Glahn in the debate on the CODA association's website — see Web Resources in Appendix B):
"J. Aitchison (2003) stated that the ilr treatment, being mathematically sound, was generally not necessary or even useless. In certain cases, he proposed the use of amalgamations and the associated log-contrasts as a more intuitive and practical way of dealing with those problems... He said: 'My complaint is not that such structure (referred to ilr and orthonormal basis in the simplex) is unimportant, but that we must not let pure mathematical ideas drive us into making statistical modelling more complicated than it is necessary.' "

The above authors counter this with the following, saying we should not move away from the "Aitchison geometry":
"...when working with compositional data which are assumed to be adapted to the Aitchison geometry of the simplex, the (apparent) simplicity of amalgamation hides its non-linear character in that geometry."

So what is the purported benefit of ILRs over simple logratios involving amalgamations? If one performs a logratio analysis (LRA) on the data, one obtains linear combinations of variables that successively maximize the explained variance. The coefficients of these linear combinations are real numbers, and in psychology, for example, rotations of solutions are routinely performed to produce more interpretable coefficients that concentrate their values into a few variables. The rotation is simple to implement, the dimensions are more interpretable, and if a two-dimensional solution is deemed adequate, the rotation results in no loss of explained variance.

ILR "balances" have a similar goal, where simple coefficients, either 1, -1 or 0 are sought, but these balances are neither simple to implement, nor easily interpretable, nor do they preserve the maximum variance in the chosen solution of reduced dimensionality. They are often used as new variables in a regression or to study interrelationships with other variables, although it makes no sense to report their summary values. While they might have some functionality as a complete set, where they reproduce the exact Aitchison geometry, using one or a few ILRs in isolation means obtaining results with variables that are difficult to interpret.

To construct a set of optimal ILR balances requires a huge amount of computational effort. In a recent article, Martín-Fernandez et al.[1] state that the algorithm is

---

[1] Martín-Fernandez JA, Pawlowsky-Glahn V, Egozcue JJ and Tolosana-Delgado R (2017) Advances in principal balances for compositional data. Math Geosc DOI:10.1007/s11004-017-9712-z

feasible for up to 15-part compositions. The first two "principal balances" of their data set, used as dimensions, account for 78.3% of the total variance, whereas the LRA explains 90.3% in two dimensions. But the most serious aspect of all this is that easily interpretable variables, simple logratios, have been replaced by logratios of geometric means, which are very complicated concepts. ILR balances are not, as Pawlowsky-Glahn et al. (2015) say, "easily interpreted in terms of grouped parts of the composition." Most people would misinterpret them as the name suggests, as amalgamated groupings of one set of parts versus another, but the term "balance" is a misnomer and counter-examples can easily be constructed (see Fig. 3.3).

Logratios using amalgamations

Using sums of parts (amalgamations) in a logratio is worthier of the term "balance". The criticism that these do not strictly obey the Aitchison geometry can be challenged. Researchers are used to variable transformations, linear as well as nonlinear ones, in an effort to arrive at models that make substantive sense, and using amalgamations in logratios rather than geometric means is a good example. I have shown several applications where using amalgamations leads to a very close approximation of the Aitchison geometry, if that is the accepted gold standard. The ability of any set of variables to reconstruct the logratio geometry can be easily measured, and experience with several data sets shows that this approximation holds. And if I record the proportions of all the alcoholic drinks that I imbibe in a month, and I want to group whiskey, vodka and aquavit into a category called "spirits", I will amalgamate their proportions by summing them, rather than the illogical operation of calculating their geometric mean — unless I've had too much to drink!

The issue of weighting

As mentioned in the Preface, Paul Lewi came up with the idea of weighting the parts in logratio analysis (alias spectral mapping) as early as the mid 1970s. Since compositional parts have differing relative errors, this idea makes a lot of sense. Fig. 6.1 is the best example in my experience of the danger of not weighting the parts. By default, the part means are used as weights, but the weights can be chosen using any justifiable criteria. Not weighting the parts, that is giving each part the same weight, can lead to serious problems when the compositions are analysed jointly, and give some parts undeserved prominence in the results. Weighting is relevant to the issue of standardization of compositional data and should always be considered.

A farewell remark

I have addressed this introductory book primarily to practitioners of compositional data analysis, who want to get the most out of their carefully collected data sets, and who want clear and understandable answers to their practical problems. I have tried to keep the arguments and methods as simple as possible, as the song goes (with apologies to the Gershwin brothers):

> *It's CODA-time, compositional data analysis is easy,*
> *Logratios are fitting, explained variance is high.*
> *Your data are rich, amalgamations good-looking,*
> *So hush, statisticians, don't you cry...*

Index